Maintenance of Historic Buildings:

A Practical Handbook

Jürgen Klemisch

Maintenance of Historic Buildings

A Practical Handbook

Jürgen Klemisch

Translated from the German by David Wade

© Donhead Publishing and Fraunhofer IRB Verlag 2011

Jürgen Klemisch has asserted his right under the Copyright, Designs and Patents Act, 1988,
to be identified as Author of this Work.

Simultaneously published in the United Kingdom and Michigan, USA by Donhead Publishing Ltd.

First published in 2011 by

Donhead Publishing Ltd
Lower Coombe
Donhead St Mary
Shaftesbury
Dorset
SP7 9LY
Tel: + 44 (0) 1747 828422
www.donhead.com

Fraunhofer IRB Verlag
Nobelstr. 12
70569 Stuttgart
Germany
Tel: +49 711 970 2500
www.baufachinformation.de

ISBN: 978 1 873394 92 2

ISBN: 978 3 8167 8002 1

Typesetting: Frauke Renz

British Library Cataloguing in Publication Data

Klemisch, Jürgen.
Maintenance of historic buildings: a practical handbook.
1. Historic buildings—Maintenance and repair.
I. Title
721'.0288-dc22

ISBN-13: 9781873394922

Library of Congress Cataloging-in-Publication Data

Klemisch, Jürgen.
[Bauunterhaltung. English]
Maintenance of Historic Buildings: A Practical Handbook / Jürgen Klemisch; Translated from the German by
David Wade; Idea and concept development, Dipl. Ing. Jürgen Klemisch (Brandenburgische Schlösser GmbH);
Team, Prof. Renate Abelmann (Büro Abelmann + Vielain Architekten) [and three others].
pages cm
"Originally published in Germany © 2006 by Fraunhofer IRB Verlag."
Includes bibliographical references.
ISBN 978-1-873394-92-2
1. Dwellings—Maintenance and repair—Handbooks, manuals, etc. 2. Historic buildings—Maintenance and
repair—Handbooks, manuals, etc. I. Abelmann, Renate. II. Brandenburgische Schlösser GmbH. III. Title.

TH4817.K5913 2011
690'.24—dc22
2010047544

Contents

Preface

This book by Jürgen Klemisch and his team exemplifies their many years of experience in the building profession. Through their expert understanding of the needs of historic buildings they have written an invaluable aid to best practice maintenance which is presented in a straightforward logical format. Although the examples provided are intended for polite architecture, the principles can be applied to all.

All too often, neglect is caused by a lack of understanding of the maintenance needs of a building – a fact long championed in the UK by the Society for the Preservation of Ancient Buildings and other key heritage organizations, and now accepted in any work centred on life-cycle costings. This excellent publication builds upon the work of Monumentenwacht in the Netherlands, widely accepted as a best practice approach

The informed approach contained in this book was developed after the reunification of Germany in 1989. This was partly as a response to the demands of an extensive works programme for many historically and culturally important castles which had been largely neglected in East Germany. The methodology emphasises the need for the long-term preservation of a building once it has been refurbished. This is achieved by setting out responsibilities and providing guidelines in a clear and user-friendly format.

By following the process outlined, a definitive logbook can be produced which covers all maintenance needs, including timing for repairs and accurate budgeting. The standard documentation is easy to understand and use, whether by contractors, service providers, caretakers or users/tenants.

The book itself is divided into two core parts plus examples from an actual project:
1. User Maintenance Instructions for use by the tenant
2. Condition Surveys for use by the owner or manager

Part 1 sets out responsibilities: what the tasks are, how they should be performed, where they are and who should undertake them. It provides simple work cards for each self-contained subject area. The need for location plans and task descriptions is set out, providing an overview in a tabulated format that is detailed yet easy to use, and which can also be updated. Specialist tasks are also identified, such as the maintenance of ventilation systems, fire alarms, water treatment and others.

Part 2 shows how routine inspections can be undertaken with minimal disturbance by the owner/manager, using a logical walk-through sequence with target times for identifying and analysing any defects. Checklists are provided throughout to ensure that all considerations are met, along with useful example spreadsheets which demonstrate how straightforward and effective the approach can be. These provide a highly effective overall framework – this attention to detail acts as a record of the issues and a simple specification.

Guardians of historic buildings, whether short-term tenants, statutory agencies or charitable organizations, should be conscious of their responsibilities and will be reassured by the expertise they now have access to in this book. Its straightforward approach will be of value to all building consultants (however experienced) and adoption of this model will be of significant benefit to the heritage sector.

Rory Cullen
Head of Buildings, The National Trust

February 2011

1   Introduction

It is an established fact that constructed entities have a limited lifespan and that the precise length of this lifespan is largely dictated by the manner in which they are treated. Any building owner should have an inherent interest in maximizing this service life through optimization of the building maintenance process A basic prerequisite of effective building maintenance is recognition of the factors which shorten the working life of a building so that their impact can be minimized. These factors include:

- Vacancies
- Neglected maintenance
- Wear and tear
- Lack of, or wrongly, implemented care measures
- Incorrect ventilation[1]
- Dilapidation due to limited lifespan of incorporated materials[2]
- Environmental action
- Infestation / infection by verm n, fungi etc.
- Building defects
- External decay agents

The experience gathered by Brandenburgische Schlösser GmbH (BSG) in the course of its work on castles, palaces, manor houses and stately homes in the German Federal State of Branden-burg has shown now easily properties can run to ruin in only a few years, particularly when unoccupied. Many of the buildings refurbished by BSG since 1993 were in a desolate state brought about by combinations of virtually all the above-named factors.

The initial step taken in every project was to avert irreversible dereliction by means of 'first-aid' repairs. Following a full-scale overhaul and occupancy-neutral refurbishment, several buildings have now been returned to full use. In these cases, the focus has shifted to preservation, i. e. how best to conserve the achieved condition in the long term.

This prompted BSG, as early as 1998, to set up a 'Building Maintenance Logbook' working group comprising four architects: Stefan Kamer, Jürgen Klemisch (both BSG), Prof. Renate Abelmann (Büro Abelmann + Vielain Architekter) and Frank Augustin (Architekturbüro Augustin). This publication presents the conclusions reached at the working group meetings and some of the initial results gathered through their implementation. To illustrate the methodology, a set of model work cards was prepared and a mode maintenance logbook compiled in collaboration with Peter Esch (gibbins european architects) using a refurbishment project handled by BSG as a basis. The author wishes to take this opportunity to thank all those involved for their whole-hearted commitment to the project.

Our first task, naturally enough, was to search for existing efficient and workable building maintenance models. While our enquiries among banks, insurers, housing associations and housing managers, and consultations with UNESCO confirmed the widespread familiarity of – and acute concern over – the underlying issues, the measures taken in this area tend to be random and uncoordinated. The UK's National Trust[3] was the only organization to provide comprehensive feedback and its stimulating input on a range of building maintenance issues led to a critical review of our previous work. The existing German standards, statutory guidelines and forms of contract have a mechanical engineering bias and are not readily applicable to buildings.[4,5,6,7] Other sources and classifications treat the subject more from a management perspective, with a focus on depreciation.[8,9,10] Workable guidelines on good building maintenance practice are lacking. At the same time, maintenance has become a burning issue in Germany, albeit subsumed under the buzzword 'facility management'. The similarly named journal[11] regularly features interesting examples of building maintenance and practical advice.[12] Symptomatic of growing awareness in this area are, for example, the revised edition of DIN 4426,[13] the information published by the Lower Saxon State Office for Architectural Conservation on the care and maintenance of cultural monuments[14] and the proceedings of a symposium on building maintenance quality management.[15] Moreover, comprehensive treatment of the issues and possible solutions are presented in the 'Handbuch Denkmalschutz und Denkmalpflege' ('Heritage Protection and Conservation Manual').[16] Based on the specific problems encountered by BSG in its refurbishment schemes, this book sets out to make its own contribution to the subject of building maintenance.

2 Aims

The working group set out to develop a workable model to promote good-practice building maintenance. The following aims were defined:

- Medium- and long-term preservation of the building fabric as far as possible in the sound, serviceable condition achieved through full prior refurbishment
- Maximization of building lifespan
- Continuous safeguarding of the building's fitness for purpose, specifically with regard to its building services systems
- Minimization of building maintenance costs
- Minimization of management expenditure
- Clear-cut definition of responsibilities
- Creation of adaptable, open-ended basic structure that allow the application of results to different buildings and their specific problems
- Provision of good-practice guidance for owners and users (tenants) for the performance of their maintenance activities or contractual maintenance obligations
- Provision of logbook for performance of regular inspections by landlords (owners) of building condition and maintenance activities (of users)

- Clear structure and easy-to-understand contents to ensure straightforward and practical routine inspections using maintenance logbook.

3 Summary

Guided by the above aims, the project work resulted in the development of a Building Maintenance Logbook with two main parts: the User Maintenance Instructions (see Section 5) and the Condition Survey spreadsheet model for routine inspections of the building condition (see Section 6).

The User Maintenance Instructions (for owners and tenants) focus on the careful, sensitive treatment of buildings and set out detailed specifications for building maintenance procedures. The Condition Survey spreadsheet model provides owners with detailed, good-practice guidance on the preparations for, performance of and follow-up to routine building inspections. Used in tandem, the two components are geared to the long-term optimization of the fabric condition, aesthetic appeal and operational performance of the building.

Over time, it will be necessary to update the Building Maintenance Logbook as circumstances change, e.g. in the case of alterations or adaptive use. Provided such updates are properly performed, the resulting tool is certain to deliver a long-term, cost-effective solution to building maintenance.

4 Collection of Background Data

4.1 Fundamentals

In Germany, architects are generally obliged, within the scope of their normal services, to present the client with the following documents of importance for later building maintenance:

- In Workstage 5 (final proposals / production information): 'Updating of final proposals / production information during production phase'
- In Workstage 8 (construction to practical completion): 'Handover of facility, including compilation and submission of necessary documentation, e.g. operating manuals, test / inspection reports, lists of defects liability periods'
- In Workstage 9 (post-completion supervision and documentation): 'Systematic compilation of drawings and computational results for facility'.[17]

There is no contractual obligation to provide similar information in the UK, so the responsibility lies with the client to request it from the supplier. The systematic acquisition, collation and handover of the above background data lays the foundation for the preparation of a Building Maintenance Logbook. However, the procurement of background information, maintenance manuals, product and company names, supply sources etc. is always laborious and time-consuming, especially when the construction or installation date of the relevant building element is long past. Hence, the procedure described below can also serve as an aid for architects in Germany in the proper execution of their obligations under HOAI § 15, and provide a reference list for architects in the UK.

It is clearly advisable to assemble all important information for the Building Maintenance Logbook at a time when it is still fresh in the memories of the involved parties. The preparation of a 'contractor questionnaire' (as described in Section 4.2) is based on a case in which a facility undergoes full-scale refurbishment (or reconstruction). Such schemes will generate most of the data important for the subsequent preparation of the Building Maintenance Logbook. These datasets should include the following:

• Details of all products used for refurbishment
• Summaries of key information
• Technical data sheets
• Information on products used
• Details of manufacturers and fabrication
• Product certifications, type codes / designations
• Test certificates
• Details of service life
• Details of load capacity
• Details of workmanship
• Information on product manufacturing process
• Element description with clear details of location
• Present status of works

4.2   Notes on contractor questionnaire (see Section 4.3)

The questionnaire form comprises two pages. The cover sheet with questions 1–6 should be completed by the architect, the second page with questions 7–18 by the relevant contractor.

4.2.1 Architect's cover sheet

Architects are closely involved in projects of this type and are normally familiar with all the key aspects. They should therefore be entrusted with the task of specifying those items that are likely to be important for the Building Maintenance Logbook.

The two fields in the top right-hand corner of the cover sheet mean the forms can be numbered or classified which helps to ensure that forms are complete. As different architectural practices employ different coding systems, no strict requirements are placed here on the format of entries. However, the use of a sequential numbering system covering all forms based on their date of use, with suitable provision for the unambiguous allocation of the second page (contractor details), e. g. through an appended number, is recommended. The use of a standard classification system (such as the 'Standardleistungsbuch' construction specification used in Germany or the National Building Specifications used in the UK) may also be helpful.

Notes on the items to be entered:

No. 1	Trade	Use cost group headings eg BCIS (UK) or DIN 276 (Germany).
No. 2	Contractor	These details allow quick contact in the event of queries or required works.
No. 3	Details of intervention/ ocation	An exact description facilitates the later allocation of instructions in the Building Maintenance Logbook.
No. 4	Contract period	Particularly where the Building Maintenance Logbook is used over long periods, this provides a clear indication of the age or installation date of the incorporated materials. Precise details can also be helpful in resolving defects liability claims.
No. 5	Building element description	An exact description allows a better assessment of the background conditions.
No. 6	Product description	The architect precisely specifies here the product about which the contractor is required to provide details on page 2 of the form. Among other things, the requirements placed on the product in the specification/bill of quantities should be listed here.

The remaining parts of the cover sheet can be used to enter additional notes or explanatory sketches. For more extensive trades involving a wide range of issues, an appropriate quantity of numbered sheets should be appended and referenced by entries in the fields in the top right-hand corner of the cover sheet. The signature panel is necessary both to establish clear responsibilities and to check on the length of time between issue date and return of the contractor details.

4.2.2 Page 2: Contractor details

Due to a wide variety of factors, the qualitative features of the finished works often diverge from the requirements of the original specification.[18,19] The responsibility for selecting a proprietary product which meets the specification normally resides with the contractor. By completing the second page of the form, contractors are forced to provide exact details of the materials used and the associated fabrication or application methods. This provides the architect with a further opportunity to check for compliance with the specification plus a wealth of valuable information for the later building maintenance process.

Before sending the form to the contractor, the appropriate sheet number (based on the system adopted by the architect – see Section 4.2.1) and the name of the contractor supplying the data should be entered in the fields in the top right-hand corner of the sheet.

Notes on the items to be entered:

No. 7 to 10		These details allow a precise designation of the product.
No. 11	Load capacity	This is used, for example, to enter data of structural relevance, such as the loadbearing capacity of building elements etc.
No. 12	Projected lifespan	This entry indicates the time when refurbishment is likely to become necessary or when more regular inspections (beyond the normal scope) are required.
No. 13	Works performed by third party	This entry indicates whether all works were actually performed by the contractor alone. Any necessary details of where to obtain further information should be provided.
No. 14	Details on workmanship	This is used, for example, to supply details on special formulations or application / installation methods a knowledge of which will be important for any subsequent maintenance, repairs, additions or refurbishment.
No. 15	Details on care	This is used to provide subsequent users with necessary information on the use of care products, treatment intervals and special guidance on application.
No. 16	Proposed maintenance	Based on their knowledge, contractors should assist intervals the architect in assessing the vulnerability of the used product with the aim of achieving long-term, trouble-free performance (e.g. for building services systems), weather protection (e.g. paintwork or water-repellent coatings) etc.
No. 17	Technical data sheets	This serves to remind contractors of their obligation to submit all relevant technical data sheets via the architect to the client. Where necessary, these should be obtained from the manufacturer and attached.

No. 18	Further details	This is used to enter any other relevant information.
No. 19	Completed	Signature of the form establishes the contractor's responsibility for the accuracy of the contents. Moreover, the date indicates the time at which the details were provided (were the works actually completed by this date?).
No. 20	Contractor details	As soon as possible after submission, the checked contractor details should be checked by the architect for coherence of content (cf. requirements described in Section 4.2.1) and completeness (including properly designated annexes). Additional details should be requested where necessary. All items up to no. 20 relate to the collection of background data. Item no. 21 serves as a final check on proper entry in the Building Maintenance Logbook.
No. 21	Entered in Maintenance Logbook	The Building Maintenance Logbook is successively prepared during the construction period and put together as a unified document upon completion of the works. The finalized Building Maintenance Logbook is the product of a review of the collected background data by the architect (see Section 4.5). The final signature is only placed after all results have been incorporated in the Building Maintenance Logbook. The collected background data are included in the annexes to the Building Maintenance Logbook.

4.3 Contractor questionnaire

Building Maintenance Logbook, collection of background data sheet   Architect's cover

| Sheet No. |
| Number of attached |
| contractor sheets |

1   Trade

2   Contractor
    (name, address, tel. no.)

3   Details of intervention /
    location

4   Contract period

5   Building element
    description

6   Product description /
    cross-ref. to BQ item no.

Issued:

Date   Architect

Building Maintenance Logbook, collection of background data   Contractor

Sheet no.

7   Name   _____

8   Make   _____

9   Manufacturer   _____

10   Type code / designation   _____

11   Load capacity (limitations
on use?)   _____

          _____

12   Projected lifespan   _____
13   Works performed by third
party? (If so, by whom?)   _____
14   Details on workmanship
(precise description of   _____
works actually performed)
          _____

          _____

15   Details on care (cross-
references to annexes)   _____

          _____

16   Proposed maintenance
intervals   _____

17   Technical data sheets
cross-references to   _____
annexes
          _____

18   Further details   _____

          _____

19   Completed: _____   _____
          Date   Contractor's signature / stamp

20   Contractor details
checked: _____   _____
          Date   Checked by (architect)

21   Entered in Maintenance
Logbook: _____   _____
          Date   Checked by (architect)

4.4 Recommendations

- To give legal force to the request for information, completion of the questionnaire by the contractor should be included as a contractual obligation in the preamble to the specification / bill of quantities.
- The best time for issuing the questionnaire is shortly before completion of the works and submission of the final account by the contractors. It is often difficult, after final payment, to obtain complete information or any details from contractors.

4.5 Review of background data

The fact that some details about a building may only be needed years after its construction – when such information may be difficult or impossible to obtain – makes the collection of background data a crucial task. Moreover, when properly ordered and included as an annex to the Building Maintenance Logbook, the background data also offers easy access to useful information whenever repairs or occupancy-related alterations, for example, are pending. The true value of the Building Maintenance Logbook, however, results from the architect's review of the data (see Section 4.2.2, item no. 21). This process involves an assessment of the details within the overall context of the building and the selective development of a maintenance agenda, specifically through the preparation of user work cards.

5   Building Maintenance Logbook, Part 1: User Maintenance Instructions

5.1 Introduction

The first part of the Building Maintenance Logbook sets out all the tasks (together with the associated documents and information) that users are required to perform to comply with their duty-of-care responsibilities. These instructions essentially address the following basic questions:

- What task needs to be performed?
- How should the task be performed?
- Where should the task be performed?
- When should the task be performed?
- By whom should the be task be performed?

5.2 Preparation of work cards

Work cards serve to provide users with detailed instructions on the maintenance tasks to be undertaken. Each work card deals with a self-contained subject area. The precise descriptions of the maintenance tasks, together with all information necessary for their performance, paves the way for the professional planning and execution of the works. The task descriptions should also be used as binding specifications in cases, for instance, where the works are subcontracted to a third-party service provider.

The work card structure provides a template into which information can be entered. Most of the required details can be extracted from the background data collection during the architect's review process (see Section 4.5).

Each work card comprises two parts. The first part defines the work location using drawings, while the second describes the tasks and presents all information necessary for their performance.

5.2.1 Part 1 of work card: definition of work location

The first part of the work cards serves to specify the work location or locations in as concise a form as possible. It must be evident at a glance where the tasks described in the second part of the work card are to be executed (see example in Section 5.3). The drawings should enable even newly appointed service providers or specialist contractors who are unfamiliar with the building to find all relevant work locations immediately. A standard layout should be adopted for all work cards on the relevant project. This should allow presentation of the items referred to in Section 5.4 in as uniform a format as possible. Similarly, a limited set of drawing templates should be prepared to chart the work locations of the various maintenance items. Most of these can be represented by means of:

- Floor plans
- Elevations

Additional templates that may be needed to accommodate special features include:

- Roof plan
- External stairs and balustrades
- Special elements, including building services
- Supplementary details
- Fixtures and fittings
- External works

The aim should be to present all work locations relevant to a particular item on a single work card. By way of example, the following recommendations can be given for the preparation of floor plan templates:

- It is recommended that the sheet be subdivided into at least as many fields as the total number of floor levels in the building.
- The floor plans should be simplified in order to include only those details essential for a clear representation of the work location. The only information required is the designation of the floor level, the outlines of all walls, stairs and openings plus the space or room numbers, allocated on the basis of a uniform system.
- To pinpoint work locations, all spaces relevant for a particular maintenance item, e. g. all rooms containing sanitary appliances, can be clearly hatched in the drawing templates. Symbols can be used in cases where hatching alone is inadequate for defining a work location. These should be explained by a key on the same sheet.
- The work card headers should include the project designation, maintenance item and a coding field.
- More extensive projects, e. g. those embracing several buildings, annexes, wings etc., may require several work cards for the same subject. In such cases, it is convenient to group together those work locations (e. g. in the same building section) with the same maintenance cycle.

5.2.2 Part 2 of work card: task descriptions

The second part of the work cards contains a precise definition of the specific maintenance tasks together with all necessary background information.

The template header (see Section 5.3) should clearly state the essential details, the most important of which is the type of work to be performed. The header should therefore include the building element or trade on which the work card focuses. The elements for which separate work cards should be prepared are listed in Section 5.4 'Structure and contents of User Maintenance Instructions'. The element or trade designation should be identical to that in the header of the first work card section (definition of work location) and, together with the coding, link the information given in the two parts of the work card. The header also states the main activities and maintenance intervals.

Main activities include:
- Routine cleaning
- Deep cleaning
- Roof cleaning
- Inspection
- Servicing
- Stripping and repainting
- Resealing

The second part of the work card features a four-column table containing all the information, in appropriate detail, needed for a long-term maintenance regime:

- Column 1 designates the task to be performed.
- Column 2, under the heading 'Work guidance', presents all information necessary for the proper performance of the task.
- Column 3 is reserved for additional comments. These may expand on the guidance provided in the previous column or offer specific recommendations. Of particular importance here is the definition of the agent or agents that are eligible to perform the task. A basic distinction can be drawn between two main groups of tasks:
  - Tasks that can be performed by the user or its agent (e. g. caretaker, cleaning contractor).
  - Tasks requiring particular expertise that can only be carried out by specialist contractors. These include the necessary servicing of the mechanical and electrical installations at predetermined intervals.

  The tasks delegated to the user essentially relate to the general housekeeping activities needed for the routine care and upkeep of the building. Any powers or responsibilities entailed by works beyond this scope should be clearly defined in the work cards! Column 3 is also used to indicate when it is necessary to notify or involve the owner.
- Column 4 is reserved for information on the materials, care products etc. needed for the specific tasks. The details on potential material supply sources and proposals regarding suitably qualified specialist contractors are intended to facilitate execution of the tasks.

The next line in the table contains an exact definition of the work location(s), i. e. a written version of the graphic information presented in the first part of the work card ('definition of work location'). Naturally, this description must tally with the details entered in the first work card section. The 'Cross-references' field allows the author to draw attention to other tasks which overlap with or supplement the described activities. This field can also be used to enter details of information sources, as-built / as-installed drawings or other documentation.

5.3 Work card template

Work card template, part 1: definition of work location (see Section 5.2.1)

Property: **Enter property name**   **workcard**

(Relevant element) **Building element**

Sample floor plan template

Basement	**Ground floor**
	(Category / serial sheet no.) **Sheet-No. ....**
First floor	**Attic**

The sheet may, for instance, be divided up according to the number of levels or elevations

Work card template, part 2: task descriptions (see Section 5.2.2)

Main activity		**Main activity**	
Frequency:	Enter maintenance intervals	Frequency:	Enter maintenance intervals
First performed:	Enter date	First performed:	Enter date

Column 1	Column 2	Column 3	Column 4
Task	**Work guidance**	**Additional comments**	**Material description / supply source**
Enter precise designation of tasks in appropriate sequence:	Enter all information necessary for proper performance of task.	Enter supplementary information: Who will perform the works? Is it necessary to appoint a specialist contractor? Must the owner be notified?	Enter specific details of materials, care products, ancillaries etc. needed to perform task. Also enter information on potential material supply sources, naming suitably qualified specialist contractors as appropriate.
Task 1 of main activity			
Task 2 of main activity			
Task 3 of main activity			
Task 4 of main activity			
Task 5 of main activity			
Text line height should be adjusted to provide adequate space for suitably detailed task description.			
Work location	Enter written description of work locations shown in first part of work card, e.g. floor levels, space / room numbers etc.		
Cross-references	Draw attention to tasks which overlap or can be performed in conjunction with tasks described here. Enter details of information sources etc.		
Issued on:	**1st update on:** Changes:	**2nd update on:** Changes:	**3rd update on:** Changes:
Name	Name	Name	Name

Sample work card, part 1: definition of work location (see Section 5.2.1)

Property: Enter property name   **Work card**

325   Sanitary appliances

Sheet No. 6 / 1

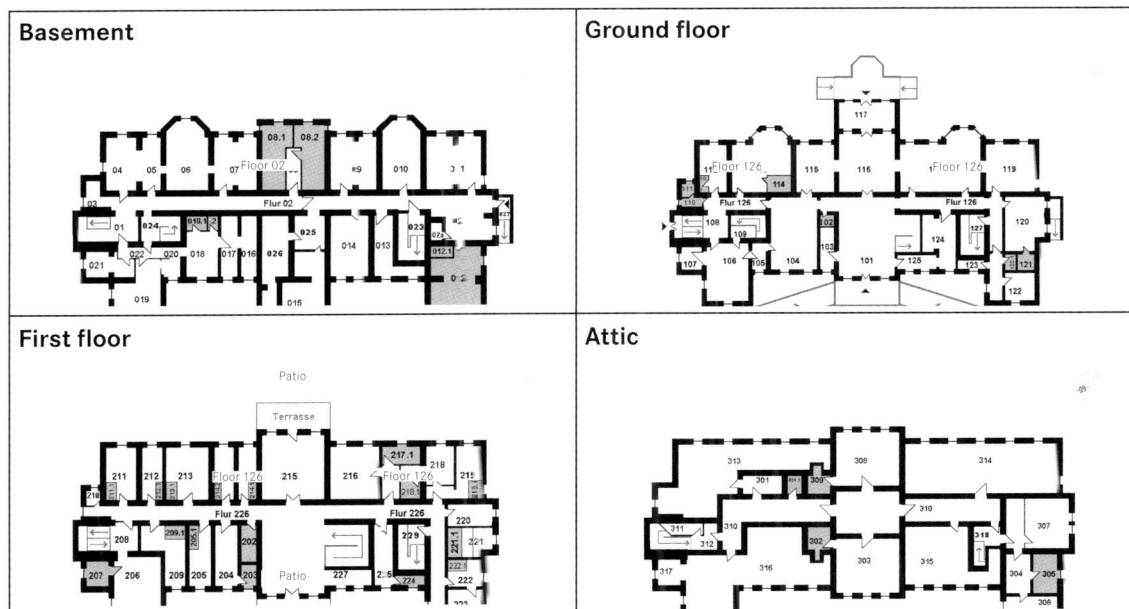

Basement

Ground floor

First floor

Attic

Sample work card, part 2: task descriptions (see Section 5.2.2)

Sanitary appliances

Inspection		Routine cleaning	
Frequency:	As required / weekly	Frequency:	As required / min. every 2 days; 2 - 5 years
First performed:	August 02	First performed:	August 02
Column 1	Column 2	Column 3	Column 4

Task	Work guidance	Additional comments	Material description / supply source
Inspection	Visually inspect for leaks, in conjunction with cleaning. 1 x annual inspection – and, where necessary, cleaning of – wash basin, shower and bathtub traps, seals to water fittings, grab rails, toilet support rails etc. (renewal approx. every 2–5 years).	Cleaning staff, caretaker Maintenance contract recommended for tiling seals.	
Silicone joint seals to showers, bathtubs	Check silicone joint seals for proper adhesion by visual and tactile inspection during routine cleaning. Renew any damaged seals.	Cleaning staff, caretaker, maintenance contract recommended, as above.	Silicone sealant
Routine cleaning of sanitary appliances	Clean with ceramic-compatible household cleaner or mild soapy water using soft cloth. Clean sanitary appliances with gentle descaler every 14 days.	Do not use abrasive or corrosive cleaners.	Standard cleaners for sanitary appliances: Villeroy & Boch "Basic" model (wall-hung WC pan, washbasin, bathtub), "Evana" model (undercounter washbasin), "Ebra" model (urinal), "Cosmos" model (shower trays)
Routine cleaning of toilet seat, flushing cistern	Clean with plastics-compatible household cleaner or mild soapy water using soft cloth. Do not allow toilet cleaners to come into contact with hinges or other fittings. When toilet cleaner is used, lift up toilet lid and seat and do not put back down until toilet cleaner has been completely flushed down WC pan.	Do not use abrasive or corrosive cleaners.	Standard cleaners
Routine cleaning of water fittings	Clean with soapy water using soft cloth or leather. Gently remove heavier soiling or scale with household vinegar or sanitary cleaner. Apply to surface, leave to act for a short time, then thoroughly rinse with clean water.	Under no circumstances should cleaners containing alcohol or gritty abrasive be used. Soft, non-abrasive sponges should also be used.	Standard cleaners for water fittings: single-lever mixer taps
Shower hair traps	Clean "from top" on daily basis or upon change of room occupancy.	Cleaning staff	
Work location	B: spaces 08.1, 08.2, 012, 012.1, 018.1, 018.2 GF: spaces 102, 110, 111, 112.1, 114, 121, 122.1 1F: spaces 202, 203, 205.1, 207, 209.1, 211.1–2 4.2, 217.1–219.1, 221.1, 222.1, 224 A: spaces 301.1, 302, 305, 309		
Cross-references	Tiling to sanitary areas, sheet no. 5 / 1–5 / 2		

Issued on:	1st update on: 17.10.2007	2nd update on:	3rd update on:
11.09.07	Changes:	Changes:	Changes:
Name Frisch	Name Frisch	Name	Name

5.4 Structure and Contents of User Maintenance Instructions

Different tasks require the use of various materials, care products, cleaning equipment, apparatus, hand tools and protective clothing. It is therefore useful to arrange the User Instructions by building element so that inspection, care and upkeep activities can be performed as a continuous operation. Each work card is thus devoted to a separate, self-contained area of maintenance.

The layout and contents of work cards are described in detail in Sections 5.2 and 5.3 respectively. An exhaustive set of examples is presented in Section 7. For this reason, the following sections include no further discussion of individual work cards, but merely list the various items for each of which a property-specific work card should, where applicable, be created:

5.4.1 Flooring

Draw up separate work cards for each type of flooring:
- Screed
- Tiling
- Marble
- Natural stone
- Terracotta
- Wood-board
- Parquet
- Linoleum
- PVC
- Carpeting
- Other

Where necessary, additional details should be provided for the treatment of skirting boards or coverings.

5.4.2 Stairs

Draw up separate work cards for each type of stair covering, as described in Section 5.4.1. Any additional details regarding the following should be integrated in the work cards or, where necessary, presented on a separate work card:
- Structural elements
- Balustrades
- Handrails
- Coatings

5.4.3 Windows and glazing

Draw up separate work cards for different construction, material and glazing types:
- Single windows
- Coupled windows
- Double windows
- Insulating-glass windows
- Timber windows
- Steel windows
- Basement windows subject to splashing
- Internal windows
- Leaded lights
- Other special glazing

The following items should also be specifically addressed:
- Coating systems
- Windows subject to heavy weathering, west elevations
- Putty / mastic joints
- Window grilles
- External-hinged shutters
- Sunshading devices
- Roller shutters
- Internal glare-control screens
- Window sills
- Ironmongery and all moving parts

5.4.4 Doors

Draw up separate work cards for each type of door or door lining finish, where appropriate, by construction:
- Steel doors
- Timber doors
- Glazed doors
- Internal doors / external doors
- Ironmongery and all moving parts
- Seals / weathering / stripping

The following items should also be specifically addressed as required:

- Opaque coatings
- Clear coatings

5.4.5 Tiling to sanitary areas

Draw up separate work cards for particular tiling features and duty:
- Floor tiles
- Wall tiles

The following items should also be specifically addressed as required:
- Special tiles
- Permanently elastic joint seals
- Access covers
- Junctions with sanitary appliances and water fittings
- Floor drains
- Junctions with other finishes

5.4.6 Sanitary appliances

Draw up separate work cards for different surface types and their functions:
- Ceramics
- Marble
- Earthenware etc.
- WC suites

The following items should also be specifically addressed as required:
- Mirrors
- (Bathroom) tops
- Holders / supports
- Shower enclosures
- Water fittings
- Other functional components

5.4.7 Electrical installations, lighting

Draw up separate work cards, by luminaire type, for those installations not serviced by a maintenance contractor:
- Internal / external luminaires
- Escape lighting
- Summary of particular system requirements for maintenance contractor

The following items should also be specifically addressed as required:
- Lamps (light sources)
- Suspension devices

- Lampshades
- Glass types, crystal chandeliers etc.
- Control buttons, sensors
- Light-level sensors
- Wet areas

5.4.8 Heating system, heat emitters

Draw up separate work cards by heat emitter type for those installations not serviced by a maintenance contractor:

- Panel radiators
- Column radiators
- Tile stoves
- Summary of particular system requirements for maintenance contractor

The following items should also be specifically addressed as required:

- Coatings
- Flow and return pipes
- Radiator covers
- Thermostatic valves
- Control units
- Fireplaces

5.4.9 Special elements

Draw up separate work cards to meet special requirements, for example:

- Balconies
- Conservatories
- Patios
- Lifts
- Cooling systems
- Swimming pools
- Saunas
- Special moveable elements, partitions and dividers
- Light wells
- Ramps
- Septic tanks
- Other

5.4.10 Fixtures and fittings

Draw up separate work cards for permanently installed items, for example:
• Wall units and coverings
• Counters
• Showcases

5.4.11 Drainage

Draw up separate work cards for:
• External drainage system
• Internal drainage system
• Summary of particular system requirements for maintenance contractor

The following items should also be specifically addressed:
· Gutters
· Downpipes
· Light wells
· Building perimeter drains, incl. associated inspection chambers and cleaning arms
· Land and surface water drains
· Fasteners for lightning protection system
· Floor drains / gullies, traps
· Strainers

5.4.12 Water, waste water

Draw up work card for:
• Waste water / sewage
• Summary of particular system requirements for maintenance contractor

The following items, among others, should also be specifically addressed:
· Grease interceptors
· Septic tanks
· Sewage lift stations
· Strainers or floor drains / gullies

5.4.13 Access covers

Draw up work card for:
• Access covers

Based on the system to be inspected, a distinction can be drawn between access covers for:

· Internal drainage pipes and stacks
· Mechanical and electrical service runs
· Chimney flues
· Cavities in walls / ceilings and behind facings
· Fire safety systems

5.4.14 Ventilation

Draw up separate work cards for:

• Ventilation
• Summary of particular system requirements for maintenance contractor

The following items, among others, should also be specifically addressed:

·   Filter replacement
·   Ventilation grilles and air filters
·   Fireplace openings
·   Housings

5.4.15 Internal walls, ceilings

Draw up separate work cards for special features, for example:

• Fibrous plaster ceilings
• Timber linings
• Textile wallcoverings

Specific items to be addressed, as required, include:

·   Fire protective coatings
·   Frescos
·   Marbling
·   Wallpaper
·   Gilding

5.4.16 External walls, façades

Draw up separate work cards for:

• Wall constructions
• Coating systems

Specific items to be addressed, as required, include:
- Plinth areas
- Cornices, string courses, corbels and other projecting features
- Subdividing / punctuating features
- Pediments
- Stucco work
- Water-repellent treatments
- Organic growth

5.4.17 Roof

Draw up separate work cards for:
- Roof areas or covering types

The following items should be addressed:
- Rain resistance
- Snow resistance
- Lightning protection system components
- Dormer window junctions
- Eaves and cornice junctions
- Chimney junctions
- Roof hatches
- Flashings
- Junctions with other structural elementsrganic growth

5.4.18 External works

Separate maintenance guidelines are currently being prepared by the author for the care of grounds, parkland and landscape architecture. In the meantime, work cards can generally be prepared in the same way as for buildings, particularly for major external structures (e. g. bridges, monuments, landscape architecture etc.).

Straightforward care and maintenance plans, including specifications, can be prepared for works to soft landscaping, water features, pathways etc. However, allowance should be made for the generally more difficult conditions governing access and site establishment. The very detailed park maintenance programmes (so-called 'Parkpflegewerke') prepared in Germany in the 1980s and 1990s have proved to be of relatively little value given the practical difficulties, arising from climatic and other non-controllable impacts, in meeting elaborate formal requirements for the external works.[20]

5.5 Maintenance contracts

A considerable number of essential maintenance tasks are not usually performed by the user and require the involvement of specialist contractors. Such works are normally performed on the basis of servicing contracts and include the maintenance of:

- Lifts and other transport systems
- Heating systems
- Electrical installations
- Ventilation systems
- Water treatment systems
- Fuel tanks
- Fire alarm systems
- Firefighting / sprinkler systems and fire extinguishers
- Burglar alarm systems
- Other protective systems

The User Maintenance Instructions should contain a list of all maintenance contracts needed. The user or owner should ensure that all such contracts are concluded promptly and the associated works carried out on schedule. The preparation of servicing plans is a very specialized task. Even at the design phase, relevant specialist engineers should give proper attention to the later mechanical and electrical (M&E) servicing requirements. Suitable provision in the tender documents issued to contractors also serves to lengthen the defects liability periods agreed for the building services trades. Given the excessive detail involved in the description of mechanical and electrical servicing plans and the ready availability of practice-oriented technical manuals in this field, no further discussion of this subject is provided here.

5.6 Work inspection cards

Work inspection cards can be used to check for compliance with the User Maintenance Instructions. They make it easier to check the status of performed and pending tasks. The inspection cards serve as checklists, presenting short-text descriptions of all the tasks described in the work cards together with the associated deadlines. It is recommended that work inspection cards are prepared in advance to cover a period of one year. All tasks (with deadlines) to be performed within this period are noted, e. g. weekly tasks will require a sheet with 52 lines.

The first header line of the inspection card is used to specify the property. As it is generally advisable to create a separate work inspection card for each particular task (deep cleaning etc), the second header line is used to enter the prescribed interval. The inspection year is entered for the user's own information and to provide an official record, as proof for the owner or landlord, of the performance of the works. The 'sheet no.' field allows sequential numbering of the inspection cards.

The six columns are reserved for the following entries:

- Column 1: projected performance date, based on specified interval
- Column 2: cross-reference to work card with long-text specification of tasks
- Column 3: basic task description in short-text form
- Column 4: agent responsible for completing task
- Column 5: actual performance date
- Column 6: signed confirmation that task has been completed

Work inspection card template

Property:_____   **Building Maintenance Logbook**   **Work inspection**

Year: _____   **Sheet no.** _____

Projected date	Work card sheet no.	Short text	Name/ contractor	Date completed	Signature
Checked for completeness:					

5.7 Data collection/work card catalogue

Given that certain tasks recur in identical or only slightly modified form on different projects, the creation of work card catalogues may be worthwhile, particularly for larger property management companies. In the same way that standard specifications are used in drafting construction contracts, such catalogues allow the fast, efficient preparation of building maintenance plans for new projects.

6 Building Maintenance Logbook, Part 2: Condition Survey spreadsheet model

The second part of the Building Maintenance Logbook is used for routine inspections of the condition of the building by the property owner or manager. It is designed to ensure that shortcomings in building maintenance are noticed promptly and that suitable remedial action takes place. This type of Condition Survey should not be confused with the detailed building surveys required prior to refurbishment schemes. The aim of the inspection procedures described here is to preserve previously refurbished building fabric through the early recognition of defects and failures.

6.1 Introduction

The following sections outline the basic preparation, execution and follow-up procedures for Condition Surveys and maintenance checks.

A major function of the Building Maintenance Logbook is to lay down guidelines and procedures for systematic and professional on-site inspections. The key instrument for this (see Section 6.3) is a series of double-page spreadsheets covering all aspects of the particular building to be examined and documented. The first (drawing) page provides a graphic definition of the relevant spaces and locations. The associated text page is used to enter a written description of any conditions diverging from the required norm. Each double-page spreadsheet deals with a self-contained topic. By turning the page, the user moves on to the next subject for inspection. The preferred layout is with the drawings positioned at the top of sheet and the texts below.

6.1.1 Structure of Building Maintenance Logbook, Part 2

Every building has its own specific demands which need to be considered during the preparation of a Maintenance Logbook. However, there are recurring issues and requirements and the methodology contained in this book aims to treat these in such a way that they can be applied to a wider range of buildings. Building-specific requirements can and should be integrated in the existing procedures in line with the adopted model.

To facilitate processing and minimize the amount of time required, the various issues are arranged into thematic groups. We recommend a breakdown that reflects the sequence of steps required for the Condition Survey. Adopting this approach, the following sections start with the preparations for the Condition Survey (Section 6.2) based on a preliminary checklist-based examination of the existing data (see Section 6.2.1.1) and a standard user questionnaire (see Section 6.2.1.2). The second stage of preparation (see Section 6.2.2) entails the analysis of the collected results, again using checklists, plus specification of the participants and date for the Condition Survey (see Section 6.3). The follow-up to the Condition Survey (see Section 6.4) focuses on defining the measures required for the long-term rectification of the identified defects and the resolution of any other pending issues.

6.1.2 Contents and page numbering

This part of the Building Maintenance Logbook should be provided with its own contents page, which also lists all related annexes together with their archiving location. Sequential page numbering is particularly important as a means of creating a self-contained log of clearly defined and recognizable scope. An example is presented in Section 8.

6.1.3 Survey intervals

Routine inspections of the building condition and checks on use-related wear and tear should be conducted at regular intervals. An annual inspection cycle is recommended to ensure the early recognition of defects. For practical reasons, these should not be governed by the official calendar year from 1 January to 31 December. Instead, a month should be specified during which the annual inspections should take place. Experience has shown the spring and autumn mid-season periods are particularly suitable given that problems tend to be most evident at these times. Best results are achieved through the proper co-ordination of maintenance and inspections. Where possible, inspections should be conducted roughly four weeks after submission of the results / reports from the annual servicing works.

6.2 Preparation of Condition Survey

Condition Surveys are normally conducted in buildings that are in service. While some disruption of building operations is inevitable, its scope and duration can be minimized and the need for additional visits avoided, with effective preparation.

A well-prepared Condition Survey will generally reduce the operational disruption and time needed on site while boosting the quality of the overall results.

6.2.1 Preliminary investigations

Preliminary investigations are a way of updating the existing stock of data. Moreover, the results may help to pinpoint any weaknesses in the contractual arrangements concluded between the user/owner and maintenance contractors. The prior identification of defects and deficiencies is crucial for the Condition Survey. Any shortcomings can be systematically checked and addressed in the course of the inspection.

The preliminary investigations comprise two steps: first, the existing data are routinely examined using a checklist (see Section 6.2.1.1). Any issues encountered here can be clarified in the second step, the follow-up user/operator questionnaire (see Section 6.2.1.2). Many of the points in the standard questionnaire form require only ticking. The inclusion of a deadline on the final page of the questionnaire is essential to ensure that the overall sequence of procedures is completed on time.

6.2.1.1   Checklist for mechanical and electrical (M&E) servicing

Preliminary investigations: servicing checklist   Code: 6.2.1.1

Element group: building services systems

Element:						
No.	**Preliminary investigations for servicing period .... to ....**	**Yes**	**No**	**If not: necessary measures**	**Initiated (date)**	**Completed (date)**
1	Has the heating system maintenance contract been submitted?			Request from user		
2	Have the results of the last scheduled heating system service check been submitted (specialist contractor's report)?			Request from user		
3	Has the sanitary/drainage system maintenance contract been submitted?			Request from user		
4	Have the results of the last scheduled sanitary/drainage system service check been submitted (specialist contractor's report)?			Request from user		
5	Has the ventilation system maintenance contract been submitted?			Request from user		
6	Have the results of the last scheduled ventilation system service check been submitted (specialist contractor's report)?			Request from user		
7	Has the electrical installation maintenance contract been submitted?			Request from user		
8	Have the results of the last scheduled electrical installation service check been submitted (specialist contractor's report)?			Request from user		
9	Has the lightning protection system maintenance contract been submitted?			Request from user		
10	Have the results of the last scheduled lightning protection system service check been submitted (specialist contractor's report)?			Request from user		
11	Has the firefighting system/equipment maintenance contract been submitted?			Request from user		
12	Have the results of the last scheduled firefighting system/equipment service check been submitted (specialist contractor's report)?			Request from user		
13	Has the fire alarm system maintenance contract been submitted?			Request from user		
14	Have the results of the last scheduled fire alarm system service check been submitted (specialist contractor's report)?			Request from user		

Property   Date: Signature:

6.2.1.2 Follow-up User/Operator questionnaire

To ....   Code: 6.2.1.2

Re: Maintenance inspection of property: ....

for period from .... to ....

Dear ....

In preparation for a forthcoming maintenance inspection, please could you submit documents with checked boxes and answer the associated questions:

Heating system
☐ Copy of existing heating system maintenance contract
☐ Copy of heating system servicing reports for period defined above
☐ Have any operational faults/problems occurred which the maintenance contractor was unable to solve? If so, what?

....
....
....
....

☐ Please state the heating energy consumption for the past calendar year.

....

Sanitary and drainage system
☐ Copy of existing sanitary/drainage system maintenance contract
☐ Copy of sanitary/drainage system servicing reports for period defined above

☐ Have any operational faults/problems occurred which the maintenance contractor was unable to solve? If so, what?

....
....
....

☐ Please state the water consumption for the past calendar year.

....

Waste disposal certificates
☐ Copy of certificate for emptying of grease interceptor
☐ Copy of certificate for emptying of septic tank

Ventilation system
☐ Copy of existing ventilation system maintenance contract
☐ Copy of ventilation system servicing reports for period defined above
☐ Have any operational faults/problems occurred which the maintenance contractor was unable to solve? If so, what?

....
....
....

Electrical installation
☐ Copy of existing electrical installation maintenance contract
☐ Copy of electrical installation servicing reports for period defined above
☐ Have any operational faults/problems occurred which the maintenance contractor was unable to solve? If so, what?

....
....

☐ Please state the energy consumption for the past calendar year.
....

Lightning protection system
☐ Copy of existing lightning protection system maintenance contract
☐ Copy of lightning protection system servicing reports for period defined above

....

Firefighting system / equipment

☐ Copy of existing maintenance contract for inspection of firefighting system / equipment

☐ Copy of firefighting system / equipment servicing reports for period defined above

☐ Have any operational faults / problems occurred which the maintenance contractor was unable to solve? If so, what?

....

....

....

Fire alarm system

☐ Copy of existing maintenance contract for inspection of fire alarm system

☐ Copy of fire alarm system servicing reports for period defined above

☐ Have any operational faults / problems occurred which the maintenance contractor was unable to solve? If so, what?

....

....

....

We would be grateful if you could complete the questionnaire and return it with the requested attachments by ....

Yours sincerely

Date, signature

6.2.2 Analysis of preliminary investigations and specification of participants and date for Condition Survey

The analysis, which should be performed separately for each building services trade (i. e. M & E installation), can commence as soon as the Follow-up User/Operator questionnaires have been returned. As the issues encountered during the analysis are likely to be very specific, the following analysis tables in Sections 6.2.2.1 to 6.2.2.7 are kept general and provide for free text entries. The available servicing reports should be checked trade by trade for any discrepancies and points requiring clarification, which should then be entered in the analysis tables.

The date of the most recent servicing should be checked for compliance with the specified maintenance interval and any deviations noted. Any issues arising from the Follow-up User/Operator questionnaire (see Section 6.2.1.2) should also be entered in the tables.

A further subject warranting attention is that of system optimization. Given the rapid pace of technological development, the potential for optimizing the existing building services systems should be examined at least every five years.

The following points should be considered:
- Adequacy of existing system in meeting requirements
- Cost-effectiveness of existing system
- Ease of use and maintenance
- Remaining lifespan of existing system
- Possibility of infrastructure changes (e. g. cabling, supply with natural gas etc.)
- Changes in legal framework

With regard to any system optimization or refurbishment, the following issues should be examined:
- Technical feasibility
- Necessary interventions in building fabric
- Projected lifespan of optimized system compared to completely new system
- Viability in terms of required capital investment
- Viability in terms of required resources
- Viability in terms of running costs
- Possibility of grant aid

The documents for the Condition Survey should be compiled by trade and any missing items obtained. Finally, the technical expertise needed to resolve the various issues should be assessed and defined. Together with the invitation to attend the Condition Survey, any appointed consultant engineers should be provided with the completed analysis tables showing the results of the preliminary investigations plus a copy of the most recent servicing report. The obligation to attend the Condition Survey should be clearly stated.

6.2.2.1 Heating system

Results of preliminary investigations: analysis   Code: 6.2.2.1
Element group: building services systems

Element: heating system
Analysis (additional details to be provided on separate sheet, as required)
When was the system last serviced by a specialist contractor?   Date:
Do any points in the servicing report require clarification? If so, please state briefly:
Do any points in the details provided by the user / operator require clarification? If so, please state briefly:
Are any problems indicated by the heating energy consumption? Please state:
Should the Condition Survey also deal with any issues relating to system optimization?
Are any other documents needed for the Condition Survey? If so, please specify documents to be obtained:
Should a consultant engineer or expert be invited to attend the Condition Survey? If so, who? Define requirements:
Forwarded, with request to attend the Condition Survey, to:
on (date) .... at (time) .... for clarification of the above issues:

Property: Date: Signature:

6.2.2.2 Sanitary and drainage system

Results of preliminary investigations: analysis   Code: 6.2.2.2
Element group: building services systems

Element: sanitary and drainage system
Analysis (additional details to be provided on separate sheet, as required)
When was the system last serviced by a specialist contractor?   Date:
Do any points in the servicing report require clarification? If so, please state briefly:
Do any points in the details provided by the user / operator require clarification? If so, please state briefly:
Are any problems indicated by the water consumption? Please state:
Should the Condition Survey also deal with any issues relating to system optimization?
Are any other documents needed for the Condition Survey? If so, please specify the documents to be obtained:
Should a consultant engineer or expert be invited to attend the Condition Survey? If so, who? Define requirements:
Forwarded, with request to attend the Condition Survey, to:
on (date) .... at (time) .... for clarification of the above issues:

Property: Date: Signature:

6.2.2.3 Ventilation system

Results of preliminary investigations: analysis

Code: 6.2.2.3

Element group: building services systems

Element: ventilation system
Analysis (additional details to be provided on separate sheet, as required)
When was the system last serviced by a specialist contractor?   Date:
Do any points in the servicing report require clarification? If so, please state briefly:
Do any points in the details provided by the user / operator require clarification? If so, please state briefly:
Should the Condition Survey also deal with any issues relating to system optimization?
Are any other documents needed for the Condition Survey? If so, please specify documents to be obtained:
Should a consultant engineer or expert be invited to attend the Condition Survey? If so, who? Define requirements:
Forwarded, with request to attend the Condition Survey, to: on (date) .... at (time) .... for clarification of the above issues:

Property: Date: Signature:

6.2.2.4 Electrical installation

Results of preliminary investigations: analysis

Code: 6.2.2.4

Element group: building services systems

Element: electrical installation
Analysis (additional details to be provided on separate sheet, as required)
When was the system last serviced by a specialist contractor?   Date:
Do any points in the servicing report require clarification? If so, please state briefly:
Do any points in the details provided by the user / operator require clarification? If so, please state briefly:
Are any problems indicated by the energy consumption? Please state:
Should the Condition Survey also deal with any issues relating to system optimization?
Are any other documents needed for the Condition Survey? If so, please specify documents to be obtained:
Should a consultant engineer or expert be invited to attend the Condition Survey? If so, who? Define requirements:
Forwarded, with request to attend the Condition Survey, to: on (date) .... at (time) .... for clarification of the above issues:

Property: Date: Signature:

6.2.2.5 Lightning protection system
Results of preliminary investigations: analysis

Code: 6.2.2.5

Element group: building services systems

Element: lightning protection system
Analysis (additional details to be provided on separate sheet, as required)
When was the system last serviced by a specialist contractor?   Date:
Do any points in the servicing report require clarification? If so, please state briefly:
Should the Condition Survey also deal with any issues relating to system optimization?
Are any other documents needed for the Condition Survey? If so, please specify documents to be obtained:
Should a consultant engineer or expert be invited to attend the Condition Survey? If so, who? Define requirements:
Forwarded, with request to attend the Condition Survey, to: on (date) .... at (time) .... for clarification of the above issues:

Property: Date: Signature:

6.2.2.6 Firefighting system / equipment
Results of preliminary investigations: analysis

Code: 6.2.2.6

Element group: building services systems

Element: firefighting system / equipment
Analysis (additional details to be provided on separate sheet, as required)
When was the system last serviced by a specialist contractor?   Date:
Do any points in the servicing report require clarification? If so, please state briefly:
Do any points in the details provided by the user / operator require clarification? If so, please state briefly:
Should the Condition Survey also deal with any issues relating to system optimization?
Are any other documents needed for the Condition Survey? If so, please specify documents to be obtained:
Should a consultant engineer or expert be invited to attend the Condition Survey? If so, who? Define requirements:
Forwarded, with request to attend the Condition Survey, to: on (date) .... at (time) .... for clarification of the above issues:

Property: Date: Signature:

6.2.2.7 Fire alarm system

Results of preliminary investigations: analysis

Element group: building services systems

Code: 6.2.2.7

Element: fire alarm system
Analysis (additional details to be provided on separate sheet, as required)
When was the system last serviced by a specialist contractor?   Date:
Do any points in the servicing report require clarification? If so, please state briefly:
Do any points in the details provided by the user / operator require clarification? If so, please state briefly:
Should the Condition Survey also deal with any issues relating to system optimization?
Are any other documents needed for the Condition Survey? If so, please specify documents to be obtained:
Should a consultant engineer or expert be invited to attend the Condition Survey? If so, who? Define requirements:
Forwarded, with request to attend the Condition Survey, to: on (date) .... at (time) .... for clarification of the above issues:

Property: Date: Signature:

6.3 Condition Survey

The procedure adopted for the Condition Survey is geared to the following objectives:
- Compilation of complete record of defects
- Provision of aids to orientation so that surveys can also be conducted by specialists unfamiliar with building
- Compilation of survey results amenable to off-site analysis
- Performance of survey within a reasonable period (target: max. one day)
- Analysis of survey results as basis for specification and implementation of remedial measures to eliminate ascertained defects
- Logical sequence for building walk-through
- Avoidance, wherever possible, of multiple walk-throughs
- Minimization of disturbance to users / tenants

The proposed walk-through sequence is based on the following:
- from outside to inside
- from top to bottom
- from large-scale to small-scale.

The survey should commence outside with an inspection of the building envelope, façades and roof. It should then continue inside with an examination of the visible structural elements (roof timbers), followed by an inspection of the individual storeys, starting at the attic level and working downwards. Particular attention should be given to the vertical circulation routes and special elements, prior to a 'large-scale to small-scale' inspection of the building services systems and a consideration of the issues raised by the analysis of the preliminary investigations (see Section 6.2.2).

The survey results are recorded using the spreadsheet system described in Section 6.1. For each survey part or subject area, the first page – hereinafter called the 'drawing page' – provides a graphic definition of the inspection location and serves as an orientation aid, particularly for survey team members unfamiliar with the building. The second page ('text page') contains written descriptions of the building condition.

The following explanations are illustrated by a full set of Condition Survey spreadsheets with drawings and text pages, presented as an annex in Section 8.

Drawing page:
The drawing page pinpoints those locations relevant to a particular subject area by means of elevations, floor plans, cross-sections and special details. These drawings may also be used to pencil in additional information regarding problems which arise.

As with the work cards, the header is used for classification purposes. The relevant building element (façade, floor level, stair etc.) is specified and assigned its position within the Building Maintenance Logbook by means of a code. The footer is used to designate the property and establish responsibilities through the date and signature. The survey date and responsible officer are entered in the side panel. Space for cross-references, additional explanations etc. is also provided here.

On the drawing page, all locations where a particular damage pattern is encountered can be indicated by appropriate markings and entry of the relevant number. This procedure ensures a clear and simple allocation of the survey findings to the relevant building elements, even for third parties processing the survey at a later date or during follow-up checks after remedial works.

Text page:

The text page is used to enter a written record of any conditions that diverge from the required norm. The entries can take the form of freely formulated text or serial reference numbers.

Given that surveys typically focus on the same, recurring problem areas, a matrix listing standard items for inspection or clarification is provided for most subject areas. This matrix serves as a checklist while also allowing the entry of numeric cross-references to the drawing page and the detailed damage records on the following page(s). Each defect type is assigned a record number for entry at the relevant position in the matrix. This procedure is recommended as it means that any defect need only be described in detail once, the record number being used to indicate its location and frequency.

Text page / damage record attachment:

Attachments to the text pages allow all survey findings to be recorded in detail on site. The descriptions provided here are assigned the same record numbers used in the drawing and text page matrix. The survey findings can then be freely described in the necessary detail on as many attached pages as needed. These descriptions are entered in Column 2. Column 3 can then be used to specify – either immediately or off-site at a later date – the necessary remedial action. Column 4 (Initiated / date / signature) and Column 5 (Completed / date / signature) are used to keep a record of the remedial works and follow-up checks. As the text page / damage record attachment always takes the same form, it will be omitted from the following detailed discussion of the Condition Survey spreadsheets.

Sample drawing page

Element group: First floor plan **Code 6.3.1**

Survey date	
Officer	
Cross-references	
Scale: none	

Property *Reichenow* Date: *14.5.2008* Signature:

Sample text page

Element group: First floor plan **Code 6.3.1**

Space	Walls		Ceiling		Flooring		Doors		Windows		Fixtures		Services	Other
	Constr.	Finish	Constr.	Finish	Constr.	Finish	Constr.	Finish	Constr.	Finish	Constr.	Finish		
204						①							②	
208							③							
211.1	④⑤						①							
213.1	⑤													
214a							①		⑥					
214.1	⑤													
215 Patio														⑦
217							①		⑥	⑧				
219.1	⑤													
222										⑧				
225								⑨	⑩⑪					

Property *Reichenow* Date: *14.5.2008* Signature:

Sample text page / damage record

Element group: First floor plan   **Text page / damage record**   **Code 6.3.1**

Element				
Finding no.	Description	Necessary action	Initiated date / signature	Completed date / signature
①	Loose floorboards	Tighten floorboard screws		
②	Cracked electric socket outlet	New socket outlet		
③	Overhead door closer disengaged / fire compartement	Refix and letter to user		
④	2 cracked wall tiles over handrinse basin	Install spare tiles		
⑤	Joint seals peeling off at wall corners	Renew joints Call out maintenance contractor		
⑥	Splits in weatherboard	Check with joiner if edging strip can be fitted, otherwise replace		
⑦	Clogged waste pipe	Clean and letter to user regarding weekly inspection		
⑧	Peeling paint (only near weatherboard)	Repaint		
⑨	Door catches when opened	Trim (plane down) door leaf		
⑩	Loose ironmongery	Secure		
⑪	Window sill loose at junction with window jamb	Remove hollows and reinstate		

Property   *Reichenow*

Date: **14.5.2008**   Signature: *[signature]*

Sample text page

Element group: First floor plan   **Text page / damage record**   **Code 6.3.1**

Element				
Finding no.	Description	Necessary action	Initiated date / signature	Completed date / signature

Property   Date: Signature:

6.3.1 Façades

Drawing page: The recommended drawing types should show a single façade, façade section or, possibly, the façade outline, together with any extensions or projecting structures accessed from the outside. Where appropriate, additional details of oriel windows, balconies, patios etc. can be provided (unless these are treated as special elements in Section 6.3.7). Any ascertained defects should be entered in the drawings at the relevant locations and assigned a serial record number (see example on previous page).

Text page:

Given the multi-faceted issues relevant to the inspection of façade condition, the text page presents a checklist specifying the primary types and patterns of damage to be considered.

The checklist is structured as follows:
- Column 1: The list of typical defects meriting special attention is designed to enhance the effectiveness of the inspection process. Property-specific additions should be made.
- Column 2: This is used for entry of record numbers consistent with those used on the drawing page. These numbers allow cross-referencing to the detailed descriptions on the attached text page / damage record.
- Column 3: This primarily acts as a checklist of the text page. If nothing is found upon examination of specific issues or building elements, this can be registered by ticking the 'n.f.' (nothing found) column.

6.3.2 Roof plan

Drawing page: The roof plan should show all hips, valleys, rooftop structures and junctions with elements penetrating the roof covering or membrane, along with the rainwater drainage system. Additional details can be provided where appropriate.

Text page:
The text page matrix specifies those elements generally requiring inspection. The subdivision of each element column allows it to be used as a checklist: a tick in the first column under each heading means that the particular roof section has been inspected and no leaks discovered, i.e. it is waterproof or no defects found (n.f.).
- Column 1 Roof section: This allows identification of the inspected roof areas and allocation of any associated findings. The roof areas can be subdivided and designated during the Condition Survey or as part of the preparations for this. It is recommended that areas be subdivided in line with on-site accessibility via roof hatches, walkways, balconies etc., following advance clarification. It may be necessary to arrange for access via a lifting platform.
- Column 2 Roof covering: In addition to a visual examination from the outside, the subsequent inspection of the attic / loft space should include a check for any signs of moisture staining

and the relevant locations entered as appropriate. Users / tenants may also be able to provide valuable information in the event of damage.

- Column 3 Chimney stacks: Any signs of damage to either the construction or surfaces / finishes should be entered. The chimney penetration, i.e. the junction of the chimney stack with the plane of the roof, warrants particular attention.
- Column 4 Valleys, ridges, hips: Any signs of damage to materials, mortar beds, fixings (displaced components) or the like should be entered.
- Column 5 Dormer windows, hatches: Particular attention should be given to the areas where these elements penetrate the roof.
- Column 6 Walkways, ladders: Particular attention should be given to underfoot safety and trafficability (moss growth, soiling).
- Column 7 Roof gutters: Particular points to look out for include signs of damage to soldered joints, expansion joints and junctions with downpipes or gargoyles, clogging and soiling (leaves), and standing water (falls).
- Column 8 Lightning protection: In particular, any loose supports, clamps and terminals should be noted. Performance tests should be left to specialist contractors.
- Column 9 Other: This column is reserved for defects that cannot be allocated to any of the previous headings.

6.3.3 Attic / loft and structural system

Drawing page: A schematic drawing of the structural roof members should be provided. The use of a rafter plan with additional cross sections showing the roof truss / framing members is a possible solution. A schematic floor plan of the attic level may be additionally provided (see drawings 6.3.3 and 6.3.4.1 in Section 8).

Text page:
- Column 1: The damage type / cause column lists a series of issues warranting separate attention and systematic inspection. These relate to the structural continuity at particular joints and connections (investigation in to whether retightening of bolts or studs is required), eaves junctions (checks for signs of water ingress), fungal or insect attack (checks for signs of bore dust, deterioration, mould formation) or displaced insulation (checks on thermal bridging due to missing or dislodged insulation). Special attention should also be given to questions regarding the building services installations located in the attic / loft space, e. g. proper ventilation, valves, distribution systems etc.
- Column 2 and Column 3: As described in Section 6.3.1.

6.3.4 Floor plans

Floor plans are used to record all the general problems identified during a building walk-through. Given the separate consideration needed for the special features of attic and basement levels, these are treated separately (see above and Section 6.3.5). The walk-through should start at the top of the building, on the attic level, and end at the bottom in the basement. Separate double-page spreads (drawing and text page) should be prepared for each level.

Drawing page: The drawing page should present full floor plans or floor plan excerpts. A relatively small scale of around 1:100 appears sufficient (drawing reduced so as to fit in space provided). The entry of the space / room numbers for clear cross-referencing to the text page and information sources in the annex is particularly important. To improve legibility, all superfluous entries, such as dimension strings or unnecessary explanations, should be omitted.

Text page:
Given the broad scope of survey items in each room or space, the key points for inspection are again systematically presented in a matrix. This provides for entries relating to all visible surfaces and finishes as well as the underlying construction. As explained earlier in Section 6, the matrix is primarily used for the entry of record numbers, to allow cross-referencing to the detailed descriptions in the text page / damage record attachment.
- Column 1 'Space': The space / room numbers in the floor plan on the drawing page can be entered here for allocation purposes.
- Column 2 'Walls': This is used to record signs of damage to wall constructions and finishes.
- Column 3 'Ceilings': This is used to record signs of damage to ceiling constructions and finishes.
- Column 4 'Flooring': This is used to record signs of damage to floor constructions and the wearing surface.
- Column 5 'Doors': This is used to record signs of damage to door constructions and finishes, including ironmongery.
- Column 6 'Windows': This is used to record signs of damage to window constructions and finishes, including ironmongery, window boards, internal hinged shutters etc. Windows should be opened and checked externally, with particular attention given to weatherboards, external hinged shutters, external window sills etc.
- Column 7 'Fixtures': This is used to record signs of damage to the construction and finishes of fixtures.
- Column 8 'Services': This is used to record signs of damage to building services installations, i.e. heating, ventilation, sanitary, electrical and all other mechanical and electrical components.
- Column 9 'Other': This column is reserved for special problems that cannot be allocated to any of the previous headings. It can also be used to enter additional notes on particular features.

6.3.5 Basement, structural waterproofing

Being in direct contact with the ground, the basement is vulnerable to a variety of problems not covered by the basic matrix items described in Section 6.3.4. These points are addressed by an additional double-page spread.

Drawing page: A basement floor plan should be provided as a means of identifying the location of any ascertained damage. This should also contain details relating to structural waterproofing, structural joints and junctions between the building fabric and the adjacent ground (or moat) that are not immediately visible or evident. The inclusion of outdoor system components, e. g. inspection chambers, septic tanks, grease interceptors etc., may also be appropriate.

Text page:
- Column 1: The damage type/cause column lists a series of issues warranting separate attention and systematic inspection, e. g. with regard to structural joints, building intake (services), damp-related tide lines on walls etc.
- Column 2 and Column 3: As described in Section 6.3.1.

6.3.6 Stairs

Drawing page: Wherever possible, each double-page spread should cover all levels of a single stairway in plan and / or section and include additional explanatory details as necessary.

Text page:
Given the variety of specific items warranting attention in connection with stairs, the key points for inspection, i.e. main stair components, are again systematically presented in a matrix.

Through the use of two sub-columns for each inspection item, provision is made for entries relating to both the visible surfaces/finishes and the underlying construction. As explained at the beginning of Section 6.3, it is recommended that the matrix be primarily used for entering record numbers, to allow cross-referencing to the detailed descriptions in the text page/damage record attachment.
- Column 1 'Stair': This is used to enter an unambiguous designation of the inspected stair and stair flight, to which the details in the relevant line relate.
- Column 2 'Treads': This is used to record signs of damage to tread constructions and finishes.
- Column 4 'Landings': This is used to record signs of damage to landing constructions and finishes.
- Column 5 'Strings': This is used to record signs of damage to string constructions and finishes.
- Column 6 'Covering': This is used to record signs of damage to covering constructions and finishes.
- Column 7 'Balustrade': This is used to record signs of damage to balustrade constructions and finishes.

- Column 8 'Handrail': This is used to record signs of damage to handrails.
- Column 9 'Other': This column is reserved for special problems that cannot be allocated to any of the previous headings.

6.3.7 Special elements

The documents under this heading can be used to address any special elements or issues not covered by Sections 6.3.1 to 6.3.6. Provision of a double-page spread is recommended for each of the following, though these are examples, not an exhaustive list:

- External stairways
- Patios
- Kitchens
- Refrigerated stores
- Transport systems, lifts
- Structures housing building services installations
- Artwork

6.3.8 Building services plant and installations

During the walk-through, any visible defects in the building services systems – e.g. cracked socket outlets, cracks in sanitary appliances, loose radiator or bathroom top brackets – are entered in Column 8 ('Services') on the text page for the relevant floor plan (see Section 6.3.4).

At the same time, the problems and issues identified during preparation for the Condition Survey (see Section 6.2.2) still require clarification. Given that the subject matter is normally very specialized, the forms used for the Condition Survey provide for the freely formulated entry of inspection results. Use of the analysis of the preliminary investigations (see Sections 6.2.1.1-6.2.2.7) is essential for conducting the Condition Survey and the largest text field is reserved for addressing these points. Space is also provided for noting down any other findings.

As the survey team members may vary depending on the particular building services trade, their names should be clearly recorded in the relevant field for later reference by the responsible officer. Participants may include specialist engineers, company representatives, architects, owners, users and caretakers. The following Condition Survey forms can be used for the building services trades:

6.3.8.1   Heating system

Condition Survey   **Code: 6.3.8.1**

Element group: building services systems

Element: heating system
Participants:
On-site inspection results regarding issues for clarification, as listed in analysis of preliminary investigations (see Building Maintenance Logbook Code 6.2.2.1):
Other conspicuous features / defects:

Property: Date: Signature:

6.3.8.2   Sanitary and drainage system

Condition Survey   **Code: 6.3.8.2**

Element group: building services systems

Element: sanitary and drainage system
Participants:
On-site inspection results regarding issues for clarification, as listed in analysis of preliminary investigations (see Building Maintenance Logbook Code 6.2.2.1):
Other conspicuous features / defects:

Property: Date: Signature:

6.3.8.3 Ventilation system

Condition Survey   **Code: 6.3.8.3**

Element group: building services systems

Element: ventilation system
Participants:
On-site inspection results regarding issues for clarification, as listed in analysis of preliminary investigations (see Building Maintenance Logbook Code 6.2.2.1):
Other conspicuous features / defects:

6.3.8.4 Electrical installation

Condition Survey   **Code: 6.3.8.4**

Element group: building services systems

Element: electrical installation
Participants:
On-site inspection results regarding issues for clarification, as listed in analysis of preliminary investigations (see Building Maintenance Logbook Code 6.2.2.1):
Other conspicuous features / defects:
Property: Date: Signature:

6.3.8.5 Lightning protection system

Condition Survey   **Code: 6.3.8.5**

Element group: building services systems

Element: lightning protection system
Participants:
On-site inspection results regarding issues for clarification, as listed in analysis of preliminary investigations (see Building Maintenance Logbook Code 6.2.2.1):
Other conspicuous features / defects:

Property: Date: Signature:

6.3.8.6 Firefighting system / equipment

Condition Survey   **Code: 6.3.8.6**

Element group: building services systems

Element: firefighting system / equipment
Participants:
On-site inspection results regarding issues for clarification, as listed in analysis of preliminary investigations (see Building Maintenance Logbook Code 6.2.2.1):
Other conspicuous features / defects:

Property: Date: Signature:

6.3.8.7 Fire alarm system

Condition Survey **Code: 6.3.8.3**
Element group: building services systems

Element: fire alarm system
Participants:
On-site inspection results regarding issues for clarification, as listed in analysis of preliminary investigations (see Building Maintenance Logbook Code 6.2.2.1):
Other conspicuous features / defects:

Property: Date: Signature:

6.3.9 External works

The inspection of external works requires specialist knowledge, among other things, of soft land-scaping, paving and watercourses. Special guidance on this subject is currently being prepared by the authors (see Section 5.4.18). At the same time, the adoption of a similar procedure to that for internal works – based on a series of self-contained subject areas, each with locational information and predetermined survey items presented on a double-page spread – is possible, particularly for garden architecture. The following bullet points are not exhaustive but suggest the types of drawing and text pages that could be prepared:

- Layout drawing or layout drawing excerpts
- Tree survey plans
- Plant survey plans
- Open waters
- Pathways
- Bridges
- Pools
- Wells
- Garden architecture
- Sculpture

- Outdoor furniture
- Steps and stairways
- Enclosures
- Pergolas

6.4 Follow-up to Condition Survey

The findings of the Condition Survey should be summarized using the prepared documentation. The summary should be arranged by building trade. Specifications prepared for rectifying defects should be similarly grouped according to the party entrusted with the remedial work. Such agents may include:

- Contractors (separately by trade)
- Service providers
- Caretakers
- Users / tenants
- Architects, engineers, expert consultants for on-site clarification of unresolved issues

Provision should be made in the award of contracts or issue of instructions for a standard period during which the ascertained damage should be rectified. At the end of this period, the success of the implemented measures should be systematically checked by means of an or-site inspection. The Condition Survey may be deemed to be complete if nothing is found during this inspection.

Notes:

[1] See ARGEBAU number 17, research report 'Richtiges Heizen und Lüften in Wohnungen' ('Proper Heating and Ventilation in Dwellings'), Aachen 4/1991.

[2] See F.W. Ross, R. Brachmann, 'Ermittlung des Bauwertes von Gebäuden und des Verkehrswertes von Grundstücken' ('Determining Construction Value of Buildings and Current Market Value of Sites'), No. 28, 1997.

[3] National Trust, 'Property Record and Maintenance Manual', 6/1993.

[4] DIN 31051, 'Instandhaltung, Begriffe und Maßnahmen' ('Maintenance: terminology and measures'), Berlin 1/1985.

[5] DIN 31052, 'Instandhaltung, Inhalt und Aufbau von Instandhaltungsanleitungen' ('Maintenance: structure and contents of maintenance guidelines'), Berlin 6/1981.

[6] 'Wartung, Inspektion und damit verbundene kleine Instandsetzungsarbeiten von technischen Anlagen und Einrichtungen in öffentlichen Gebäuden' – Wartung 85 ('Servicing, inspection and associated minor repairs to building services plant and installations in public buildings' – Servicing 85), AMEV (Mechanical and Electrical Engineering Working Group of State and Municipal Administrations) in German Federal Ministry for Regional Planning, Building and Urban Development, Bonn 1993.

[7] 'Vertragsmuster für Instandhaltung (Wartung, Inspektion, Instandsetzung) von technischen Anlagen und Einrichtungen in öffentlichen Gebäuden' – Instandhaltung 90 ('Standard form of contract for maintenance (servicing, inspection, repair) of building services plant and installations in public buildings' – Maintenance 90), AMEV (Mechanical and Electrical Engineering Working Group of State and Municipal Administrations) in German Federal Ministry for Regional Planning, Building and Urban Development, Bonn 1990.

[8] DIN 18960 'Baunutzungskosten von Hochbauten, Begriffe, Kostengliederung' ('Cost-in-use of buildings: terminology, cost classification'), Berlin 4/1976. The UK equivalent is BS ISO 15656-5 Buildings & Constructed Assets – Service Life Planning.

[9] II. Berechnungsverordnung (Second Computation Ordinance), German Federal Law Gazette number 55 Z 5702 A, Bonn, 18 October 1990, amended 13 July 1992 (Gazette p. 1250), published in 'Planen und Bauen' 40/1993.

[10] 'Handbuch der kostenbewussten Bauplanung' ('Manual of Cost-Conscious Construction Design'), Wuppertal 1976.

[11] 'Facility Management', Bertelsmann Fachzeitschriften GmbH, Gütersloh.

[12] For example, Hellerforth, M. 'Optimierung gewünscht, Adäquate Einbindung des Facility Managements schon in der Neubauphase' ('Optimization required, proper integration of facility management in new-build phase'), *Facility Management*, No. 5, 2002.

[13] 'Sicherheit bei Wartung und Instandsetzung, Neufassung der DIN 4426, jetzt konkrete Planungshilfe' ('Safety in servicing and repair, new edition of DIN 4426 with practical design aid'), Torsten Genthein Deutsches Architektenblatt 4/02.

[14] Denkmalschutz Informationen (Heritage Preservation Information), published by Deutsches Nationalkomitee für Denkmalschutz (German National Committee for Preservation of Historic Monuments), Federal Government Commissioner for Culture and Media, Bonn, 02/2002 edition, year 26, pp. 38–43.

[15] 'Neue Wege zur Bauwerkserhaltung und Denkmalpflege' ('New approaches to the preservation of architectural and historic monuments'), proceedings of symposium entitled 'Qualitätsmanagement in der Bestandspflege' ('Building Maintenance Quality Management') as part of an architecture and heritage initiative in Braunschweig, Erich Schmidt Verlag GmbH & Co., Berlin, 2002.

[16] Martin, D. J. and Krautzberger, M., 'Handbuch Denkmalschutz und Denkmalpflege' ('Manual for Protection and Care of Historic Monuments'), Section D Care of Historic Monuments, IV Preservation, Precautions and Care, and IX Second Model: Bamberg Maintenance Contract, published in collaboration with Deutsche Stiftung Denkmalschutz (German Foundation for Monument Protection), Verlag C.H. Beck, Munich, 2004.

[17] HOAI (Official Scale of Fees for Services by Architects and Engineers), 5th Amendment Ordinance for Ninth Euro Introduction Law; Fee Scales, Werner, Düsseldorf, 2001.

[18] VOB (German Construction Contract Procedures) Berlin; Vienna; Zurich; Beuth 2000. The UK equivalent is the JCT Forms of Contract: there is no requirement for specifications prepared in accordance with these to be product-neutral.

[19] See VOB (German Construction Contract Procedures), Part A § 9.5.

[20] Schulz, H., 'Der Umgang mit denkmalgeschützten gärtnerischen Anlagen bei der Brandenburgischen Schlösser GmbH' ('Treatment of listed garden facilities by Brandenburgische Schlösser GmbH'), Holger Schulz, pp. 40-48, 'Denkmalpflege in Berlin und Brandenburg – Gartenkunst und Gartendenkmale', ('Heritage Conservation in Berlin and Brandenburg – Garden Art and Monuments'), Michael Imhof Verlag, Petersberg, 2004.

[21] For example, on servicing, inspection and associated minor repairs to building services plant and installations in public buildings – Wartung 85 (Servicing 85) – forms of contract, inventory, specifications, AMEV (Mechanical and Electrical Engineering Working Group of State and Municipal Administrations) in German Federal Ministry for Regional Planning, Building and Urban Development, Bonn 1993, plus standard forms of contract for maintenance (servicing, inspection, repair) of alarm systems (fire/burglar/hold-up alarms and outdoor surveillance) – Instand GMA 94 (Alarm Systems Repair 94), AMEV, Bonn, 1994.

7　Example of User Maintenance Instructions

Section 5 explains how to prepare User Maintenance Instructions. The example shown in the following pages serves to illustrate the practical application of the described method.

The Specimen documents are based on a real building, Schloss Reichenow, which is located in the German Federal State of Brandenburg and described in greater detail in Section 9.

User Maintenance Instructions are particularly crucial for the conservation of historic building fabric. A precise specification of maintenance tasks is the first step in enabling users to meet the many special requirements of such situations. Yet long-term value preservation is an issue of equal importance to all buildings and the described procedure is, in principle, applicable to any facility type – including residential and administrative properties, factories, sports centres or shopping malls – not just listed buildings. By illustrating the procedure outlined in Section 5, the following examples are designed to help readers prepare their own, property-specific User Maintenance Instructions.

Index of work cards

Ground floor

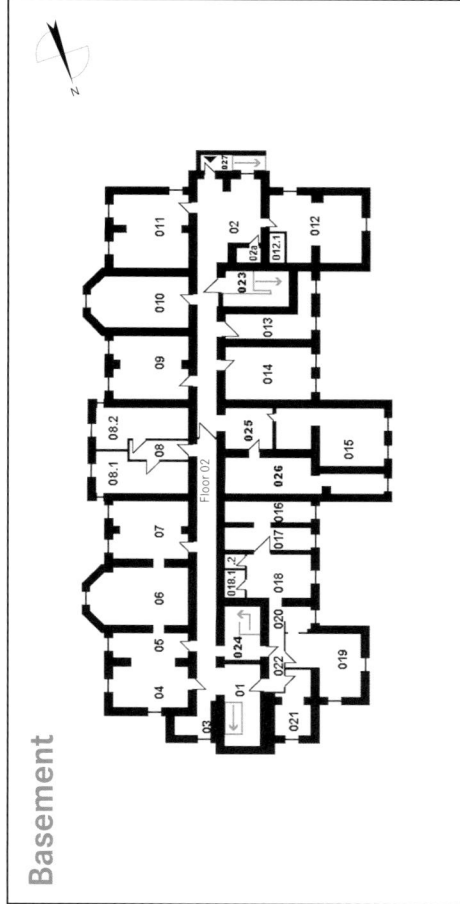

Attic

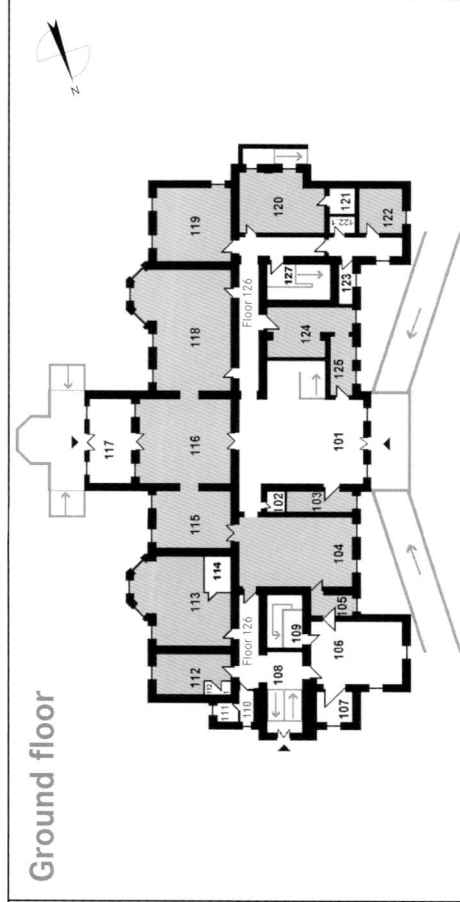

Basement

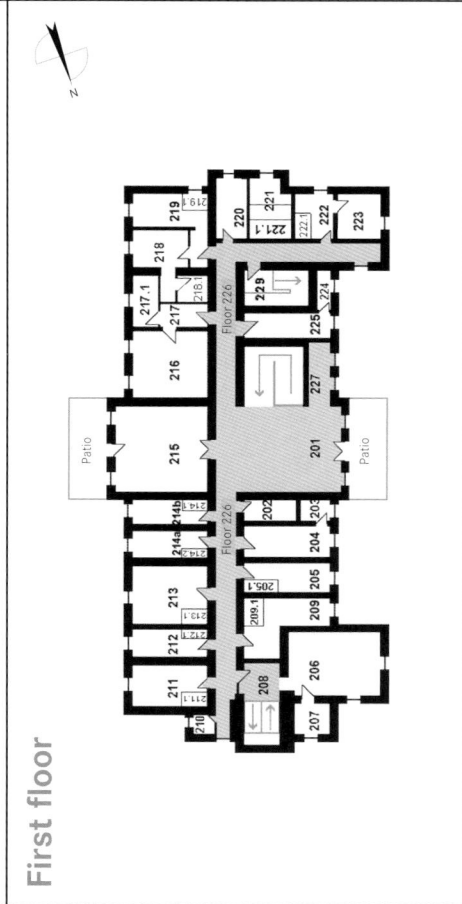

First floor

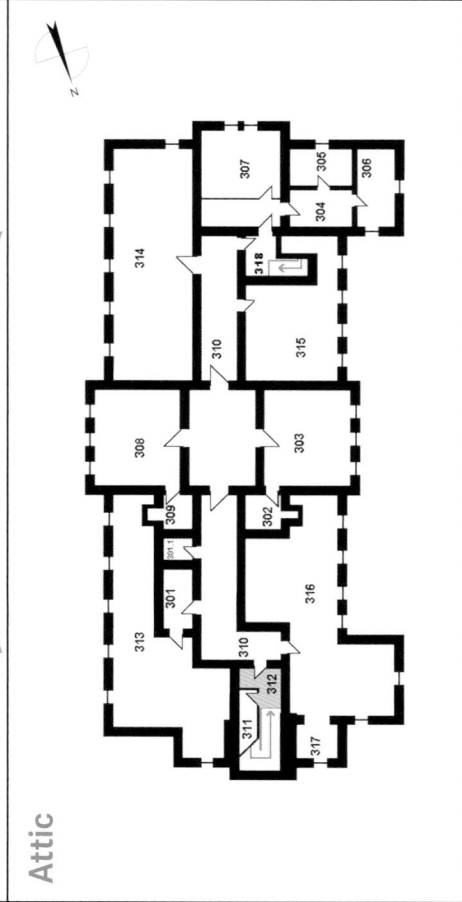

Routine cleaning		Deep cleaning	
First performed:	As required / min. every 14 days	Frequency	As required
	May 05	First performed:	May 05

Task	Work guidance	Additional comments	Material description / supply source
Routine cleaning of timber flooring	Remove rough dirt, dust down with slightly damp cloth, wipe / clean floor with slightly damp mop or cloth. Add vinegar cleaner to cleaning water. When cleaning medium- to heavy-duty floor surfaces, add L93-Everclear product to cleaning water.	Ensure required ambient temperature and humidity are maintained to avoid cracking: ideal ambient temperature: 18–20°C; ideal humidity: 50–60%.	Care system comprising L 93-Everclear and L 94-Cleaner-Wax Remover or equivalent products.
Deep cleaning of timber flooring	In case of unsightly flooring, clean surface with L 94-Cleaner. Then perform finish care treatment.	Use of humidifier is recommended in heating season.	Possible supply source: parquet merchant.
Finish care for timber flooring	A finish care treatment with undiluted product is required after deep cleaning. Perform finish care treatment with L 93-Everclear. Apply thin film in undiluted form. Gloss appears without additional polishing.		If known, enter name of contractor here with address, tel / fax, e-mail and internet address.
Removal of local soiling and stains	Wipe away dirt marks or stains with cloth using undiluted L 94-Cleaner.		
Floor sealing	**Every 10 years for timber flooring.**	Subject to prior consultation with owner. To be performed only by suitable specialist contractor.	Treatment with oil / resin-based sealer. Specialist contractor as supply source (see above).
Work location	**Ground floor: spaces 103, 104, 105, 112, 113, 115, 116, 118, 119, 120, 122, 124, 125** **First floor: spaces 201, 208, 215, 226, 227** **Attic: spaces 312**		
Cross-references			

Issued on: 02.03.2007	1st update on: 17.10.2007	2nd update on: 15.07.2008	3rd update on:
Changes:	Changes:	Changes:	Changes:
Name: Miethe, Plätzer	Name: Frisch	Name: Klemisch	Name:

Ground floor

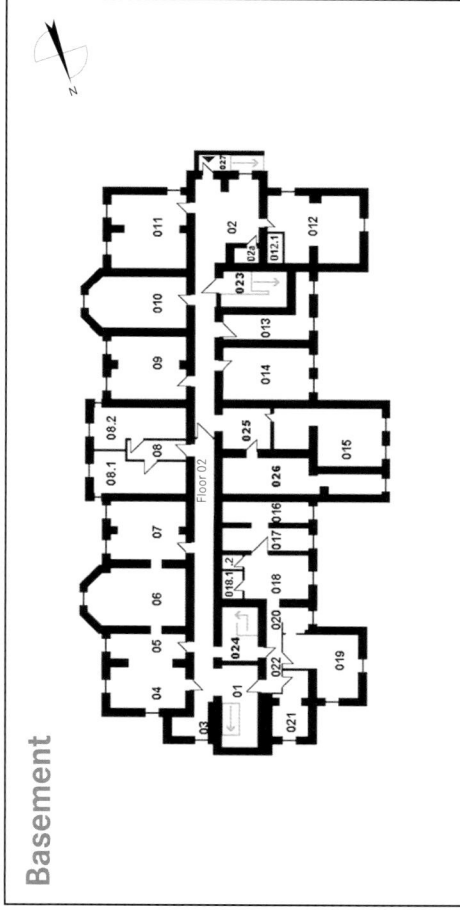

Attic

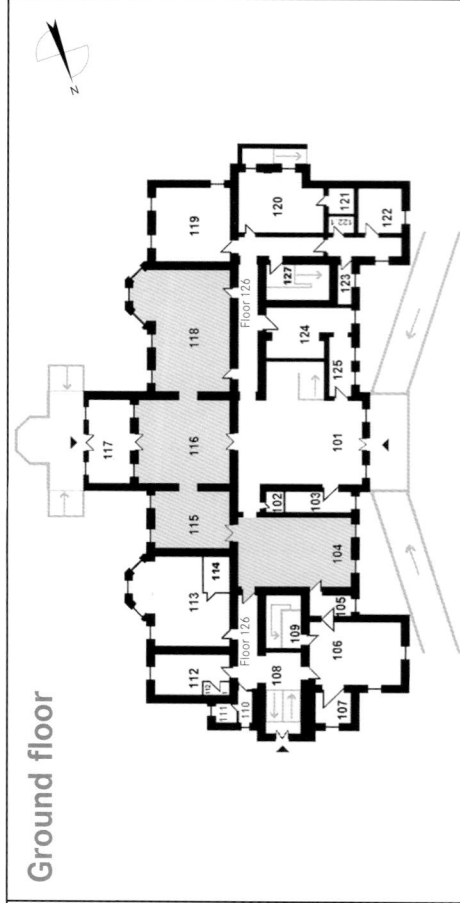

Basement

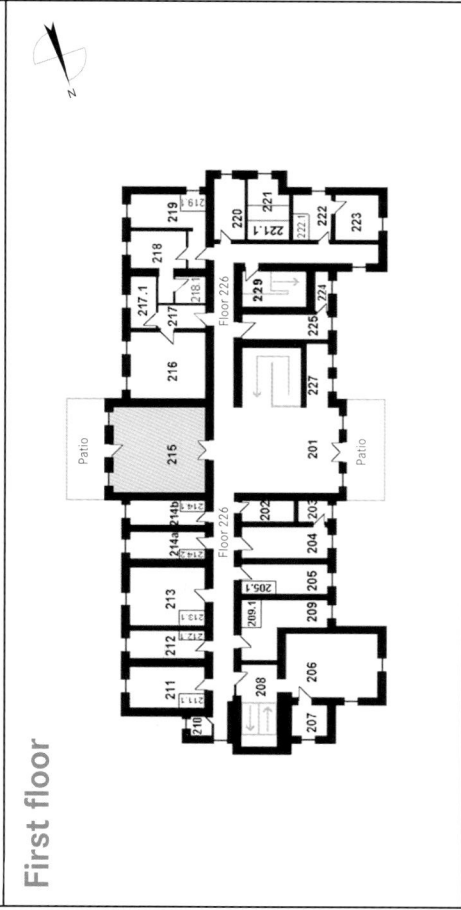

First floor

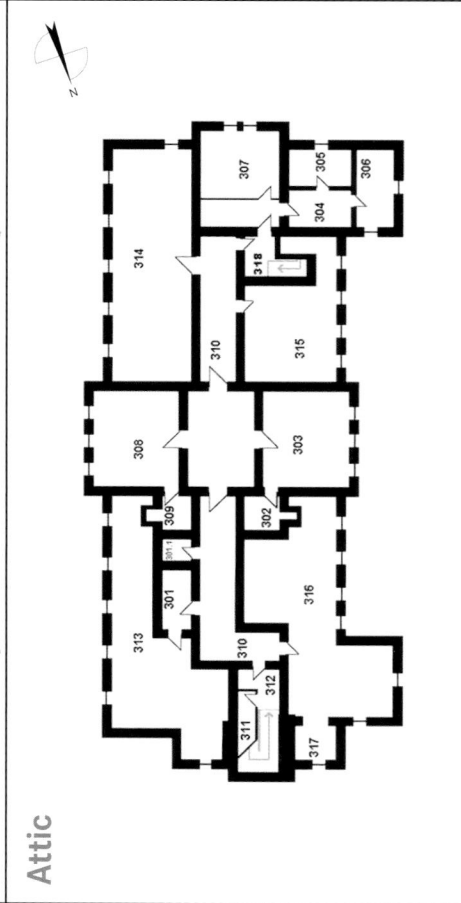

Floor sealing

Frequency:

First performed:

Grinding / replacement

Frequency: Every 15 years

First performed: May 05

Task	Work guidance	Additional comments	Material description / supply source
	Every 7 years	Frequency:	Every 15 years
	May 05	First performed:	May 05
Parquet grinding	Mechanically grind in several stages and subsequently varnish. Select grinding method and grit size as required by type of soiling and damage.	Moisture content at time of installation max. 12%. Subject to prior consultation with owner. To be performed only by suitable specialist contractor.	Wood strip: historic, seasoned pine parquet.
Parquet replacement	Base: timber joists with loam pugging, fix firrings up to 5 cm for level base and place felt strip underlay; lay seasoned, planed, tongued and grooved, 28 mm thick, 17.5 cm wide pine parquet strips, perimeter insulation strip, countersink nails and stop holes.	Use of humidifier is recommended in heating season. Subject to prior consultation with owner. To be performed only by suitable specialist contractor.	Possible supply source: parquet merchant. If known, enter name of contractor here with address, tel / fax, e-mail and internet address.
Floor sealing	With oil / resin-based sealer, allowing sufficient time for drying.	Subject to prior consultation with owner. To be performed only by suitable specialist contractor.	LT-Export-Extra, for example. Specialist contractor as supply source (see above).
Work location	Ground floor: spaces 104, 115, 116, 118 First floor: space 215		
Cross-references			
Issued on: 02.03.2007	1st update on: 09.10.2007	2nd update on: 15.07.2008	3rd update on:
Name: Miethe, Plätzer	Changes: Name: Frisch	Changes: Name: Klemisch	Changes: Name:

Property:

322

Floors /terrazzo
Sheet no. 1/5

Ground floor

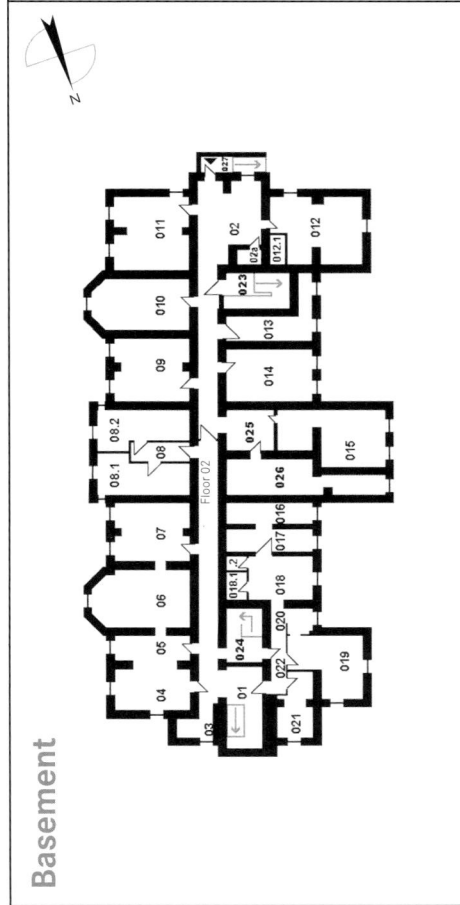

Attic

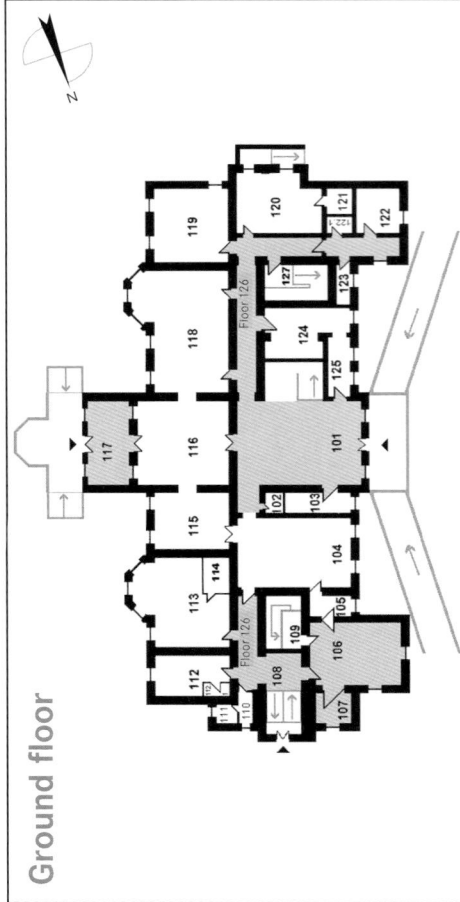

Basement

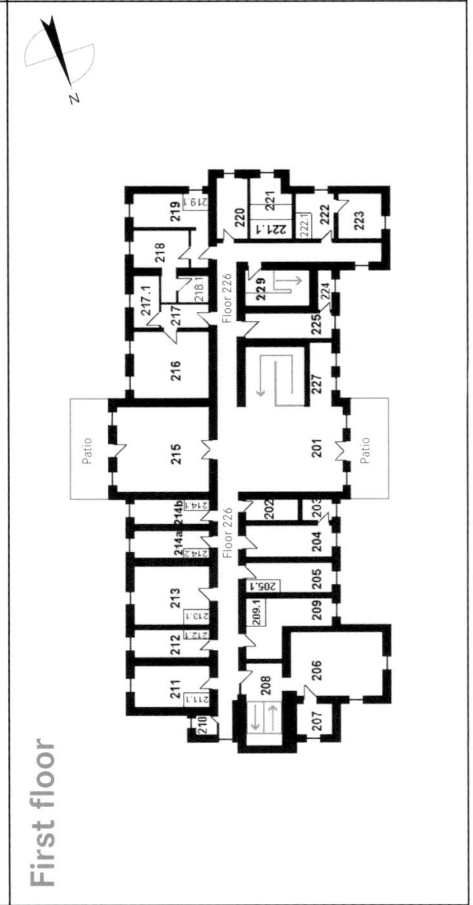

First floor

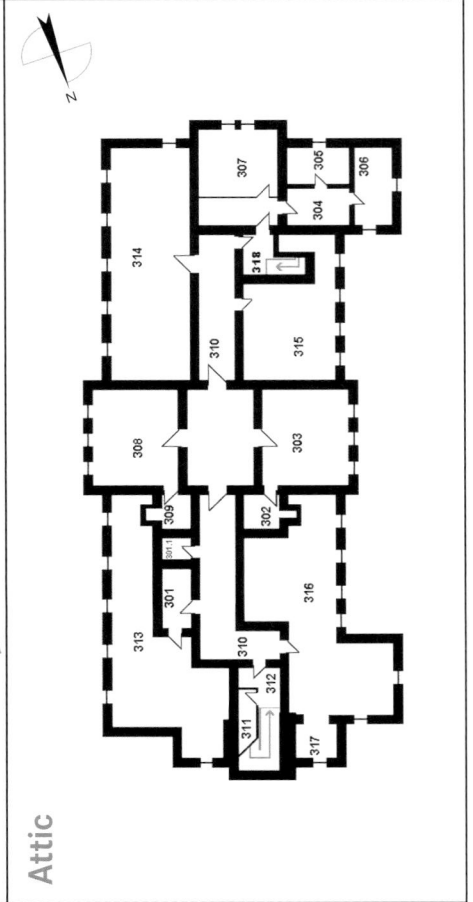

Routine cleaning		Deep cleaning	
Frequency:	As required / min. every 5 days	Frequency:	As required / min. every 14 days
First performed:	May 05	First performed:	May 05

Task	Work guidance	Additional comments	Material description / supply source
Routine cleaning of terrazzo flooring	Clean only with pH-neutral cleaners and cloth! Clean frequently with vacuum removal; mechanical cleaning with scrubbing and vacuuming in single operation is recommended for large floor areas.	Do not use sanitary or aggressive cleaners. Do not use additives such as wax etc. (impairs slip resistance). Do not use scouring agents or liquids. Cleaning aids such as Scotch-Britt scratch surface.	Recommended cleaner: standard neutral soap/neutral cleaner (soft soap) or SR 13 special cleaner. Supply source: enter address, tel/fax, e-mail and internet address.
Deep cleaning of terrazzo flooring	At regular intervals; use mechanical method to combat heavy soiling (grease, wax, oil etc.) Use alkaline deep cleaner.		Alkaline deep cleaner.
Anti-slip treatment	For data sheet on floors in workrooms and working spaces with slipping risk, see annex. For cleaning and care, see Section 5.1, pp. 13–14.	To be performed only by suitable specialist contractor.	Anda-Segura system, for example. Enter specialist contractor with address tel/fax, e-mail and internet address.

Work location	**Ground floor: spaces 101, 106, 107, 108, 117, 126**

Cross-references			
Issued on: 02.03.2007	1st update on: 05.09.2007	2nd update on: 15.07.2008	3rd update on:
Name: Miethe, Plätzer	Changes: Name: Frisch	Changes: Name: Klemisch	Changes: Name:

Sheet no. 1 / 6

Ground floor

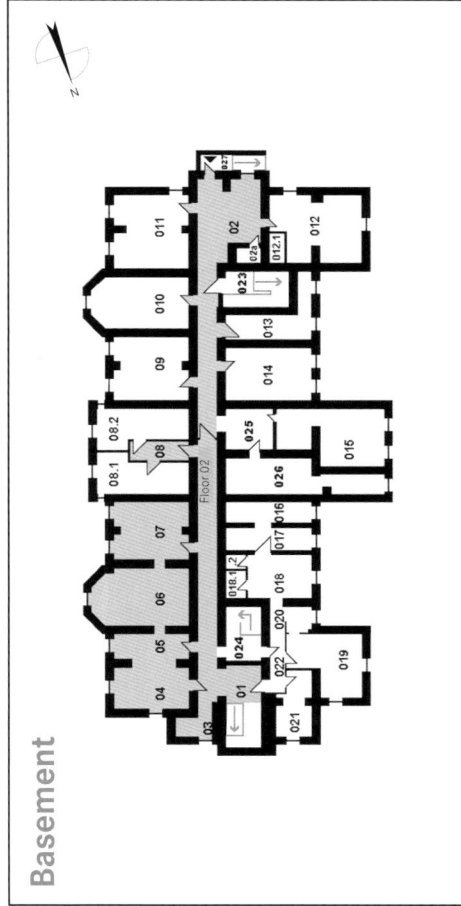

Attic

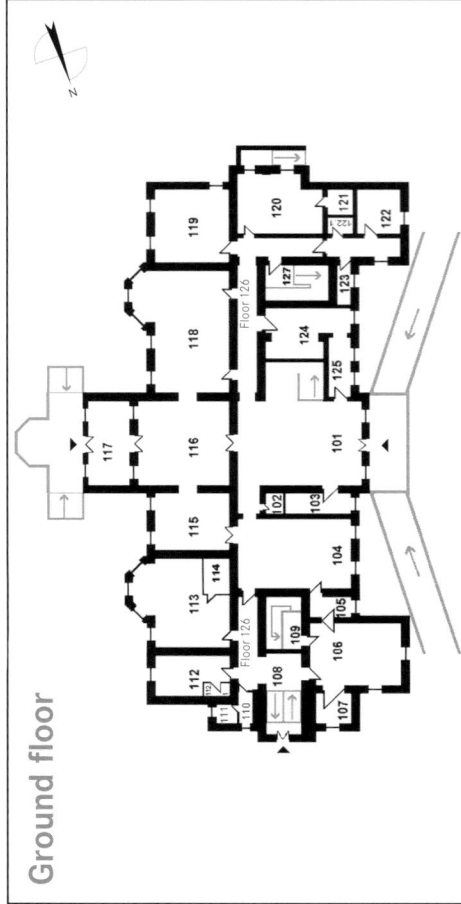

Basement

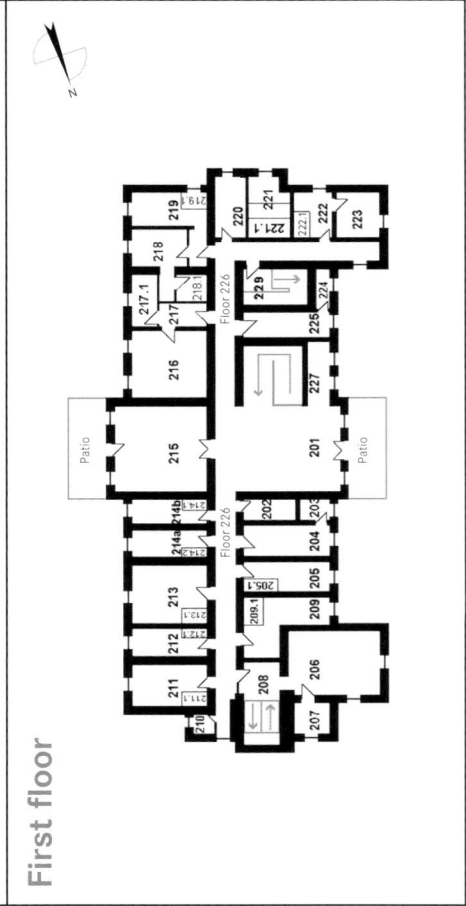

First floor

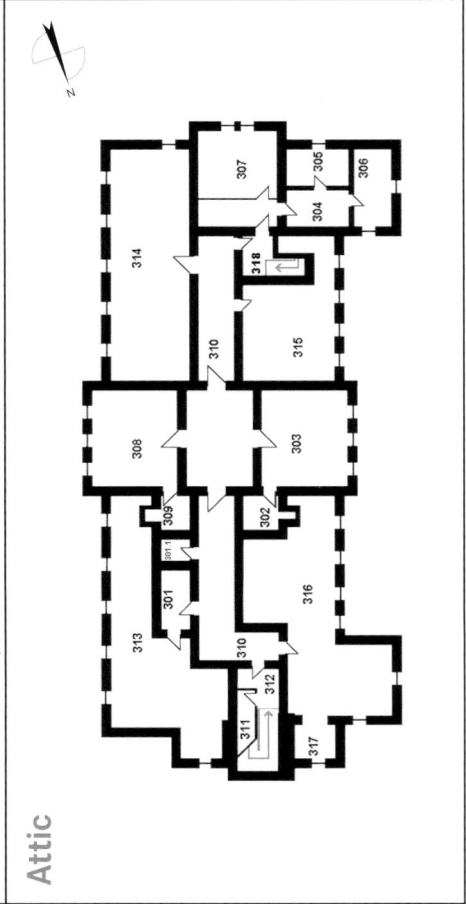

Routine cleaning	Deep cleaning
Frequency:	Frequency:
As required / min. every 5 days	As required / monthly
First performed:	First performed:
May 05	May 05

Task	Work guidance	Additional comments	Material description / supply source
Routine cleaning of ceramic flooring	Sweep with broom and wipe down with slightly damp mop or cloth. Add neutral soap to cleaning water in proportion 1:50.	Cleaners leaving film on stain-repellent surface are unsuitable. Do not use solvent-bearing or leaching cleaners.	Addition of neutral soap to cleaning water or standard proprietary care products for hard stoneware flooring are suitable. Obtainable from tile merchants. Enter proposed product. Supply source for ceramic slabs: Enter company address, tel/fax, e-mail and internet address. History series, colour Provence, size 240 x 240 x 18 mm.

Work location	Basement: spaces 01–08

Cross-references

Issued on: 02.03.2007	1st update on: 10.10.2007	2nd update on: 15.07.2008	3rd update on:
	Changes:	Changes:	Changes:
Name: Miethe, Plätzer	Name: Frisch	Name: Klemisch	Name:

Sheet no. 1 / 8

Ground floor

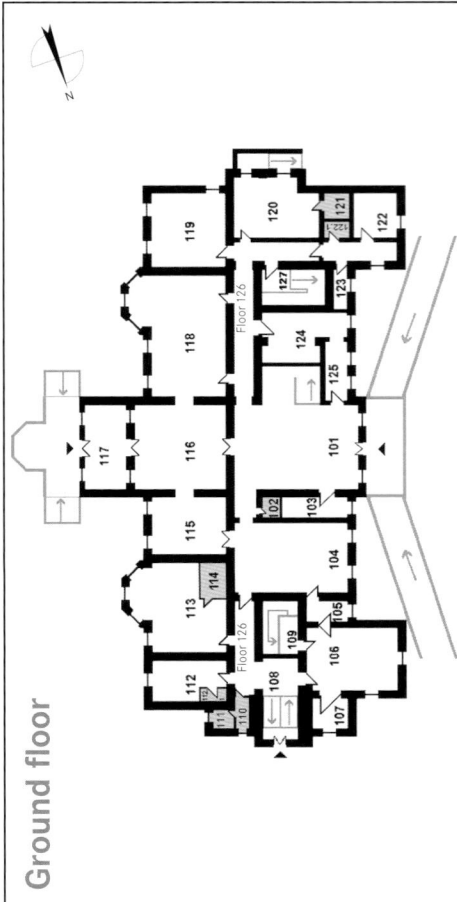

Attic

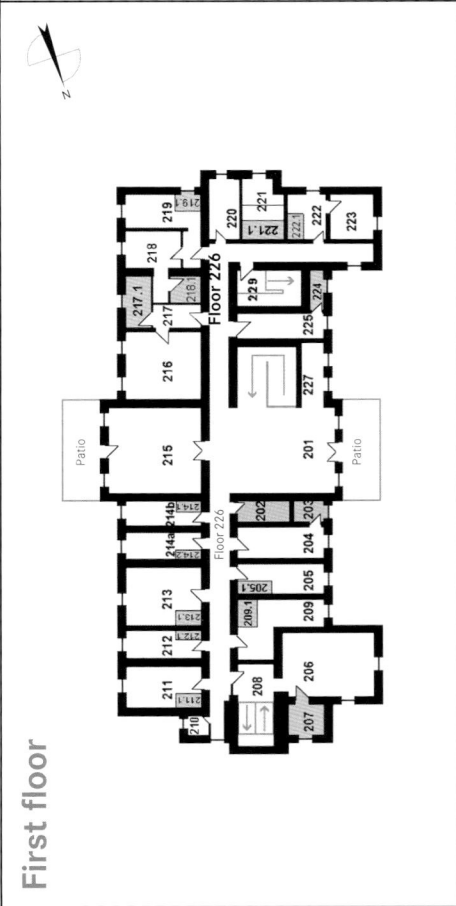

Basement

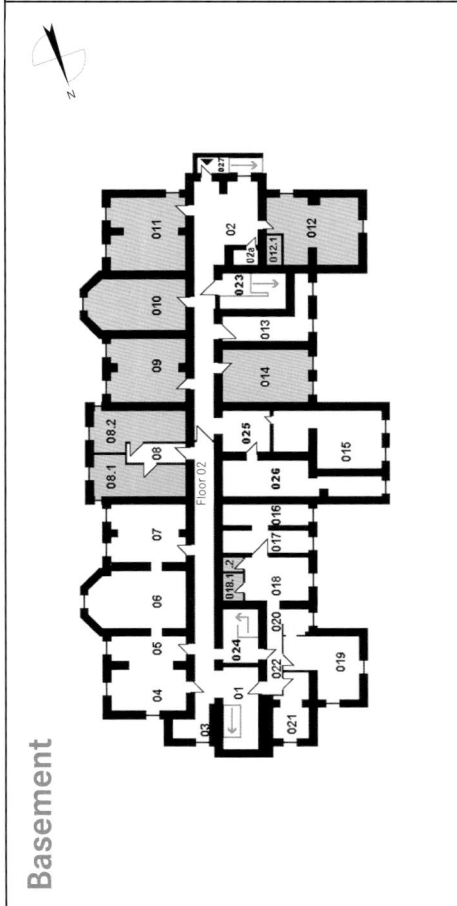

First floor

Routine cleaning		Deep cleaning	
Frequency:	As required / min. every 5 days	Frequency:	As required / monthly
First performed:	May 05	First performed:	May 05

Task	Work guidance	Additional comments	Material description / supply source
Routine cleaning of tiles	Sweep with broom and wipe down with slightly damp mop or cloth. Add tile cleaner to cleaning water. Clean tiles with descaling agent every 14 days.	Cleaning staff.	Standard proprietary tile cleaner and descaler. Supply source: enter address, tel/fax, e-mail and internet address.
Inspection	Visually inspect joint seals in conjunction with cleaning.	Cleaning staff, caretaker.	
Cleaning of floor drains	Remove dirt from strainers, every day when room is occupied. Rinse floor drains with water every month to prevent build-up of odours.		Silicone sealant.
Renewal of joint seals	To be performed when peeling and cracks are visible. In areas exposed to splashing, approx. every 2–4 years, otherwise approx. every 4–10 years.	To be performed by suitable specialist contractor.Maintenance contract with specialist contractor recommended, for periodic (half-yearly) renewal of damaged areas only.	Supply source for tiles in spaces 009, 010, 011, 014: enter company. Fully vitrified stoneware, Rustic style, Belforte Rosa tiles, size 300 x 300 mm, R 9 anti-slip, abrasion class IV. Ggf. Enter maintenance contractor. Tiling to sanitary areas: enter company. Size 200 x 200 mm.

Work location	Basement: spaces 08.1, 08.2, 09, 010, 011, 012, 012.1, 014, 018.1, 018.2
	Ground: spaces 102, 110, 111, 112.1, 114, 121, 122.1
	First floor: spaces 202, 203, 205.1, 207, 209.1, 211.1–214.2, 217.1– 219.1, 221.1, 222.1, 224 Attic: spaces 01.1,302,305,309

Cross-references	Tiling to sanitary areas, sheet no. 5 / 1–5 / 2

Issued on: 02.03.2007	1st update on: 17.10.2007	2nd update on: 15.07.2008	3rd update on:
	Changes:	Changes:	Changes:
Name: Miethe, Plätzer	Name: Frisch	Name: Klemisch	Name:

Ground floor

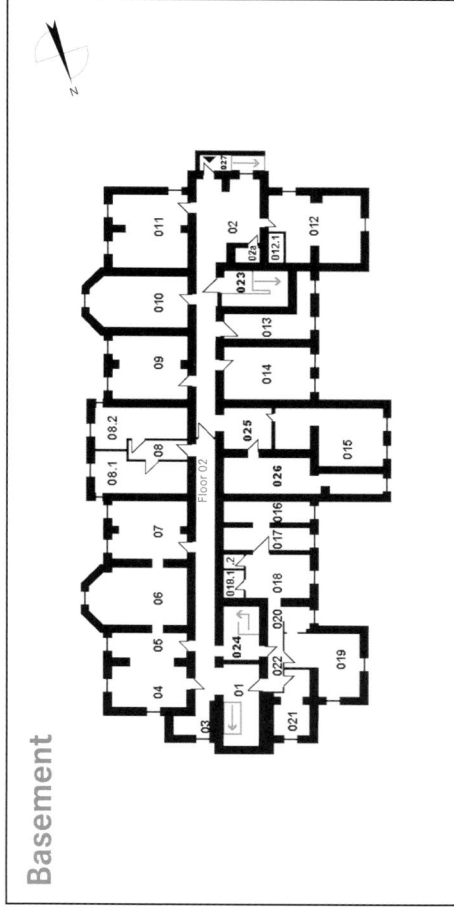

Attic

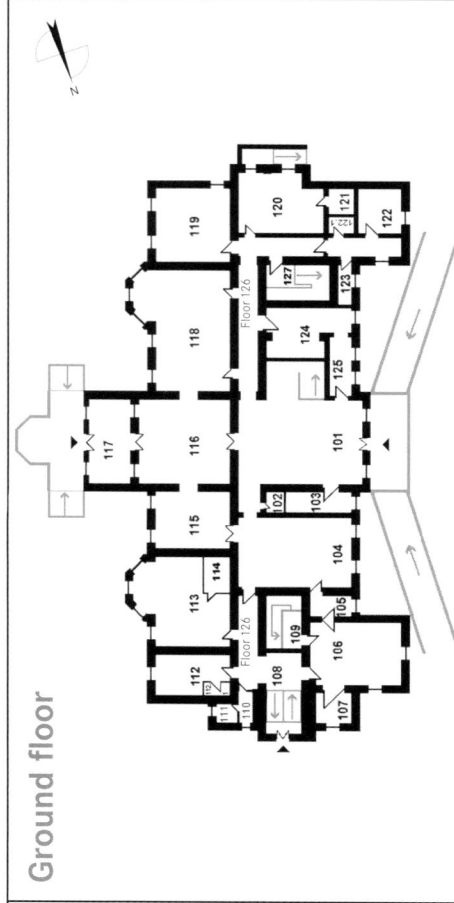

Basement

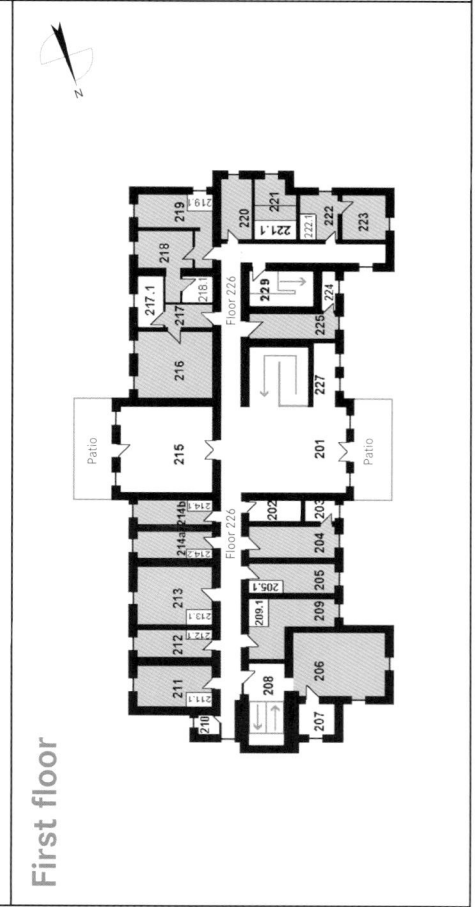

First floor

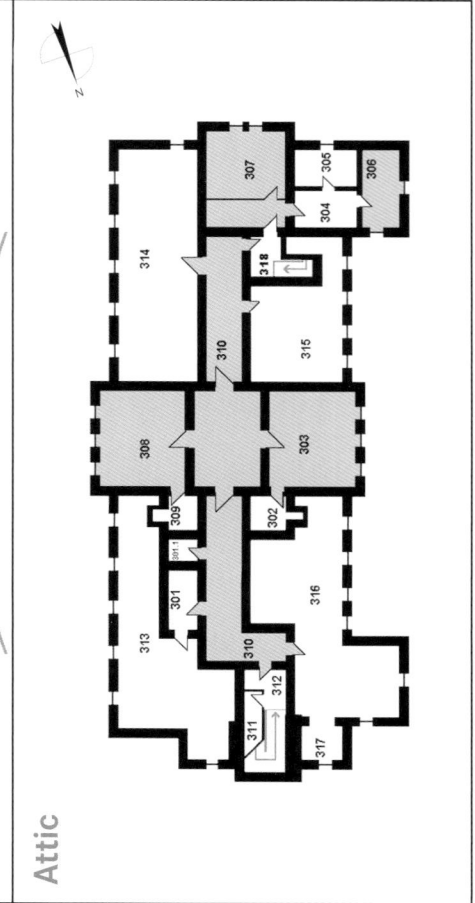

	Routine cleaning	Deep cleaning		
Frequency:	As required	Frequency:		As required
First performed:	May 05	First performed:		May 05

Task	Work guidance	Additional comments	Material description / supply source
Routine cleaning of carpeting	Vacuum heavily trafficked carpeting with powerful vacuum cleaner.		
Deep cleaning of carpeting	Using wet vacuum cleaner with machine-specific cleaners.		
	Clean metal carpet strip with slightly damp cloth, without scouring agent		Carpet supply source: enter contractor with address, tel/fax, e-mail and internet address. Quality King 1111,1000 g, design 97151-3079-2.

Work location	First floor: spaces 204–206, 209, 211–214b, 216–223, 225
	Attic: spaces 303, 306, 307, 308, 310

Cross-references			
Issued on: 02.03.2007	1st update on: 17.10.2007	2nd update on: 17.05.2008	3rd update on:
	Changes:	Changes:	Changes:
Name: Miethe, Plätzer	Name: Frisch	Name: Klemisch	Name:

Ground floor

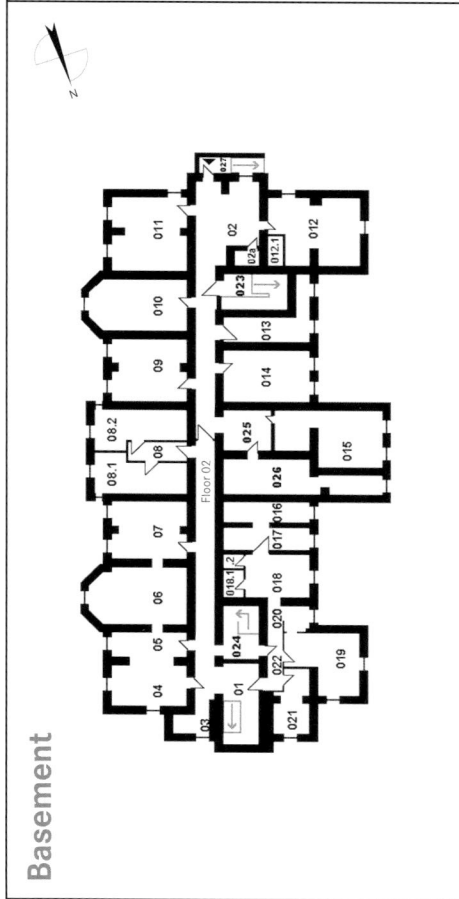

Attic

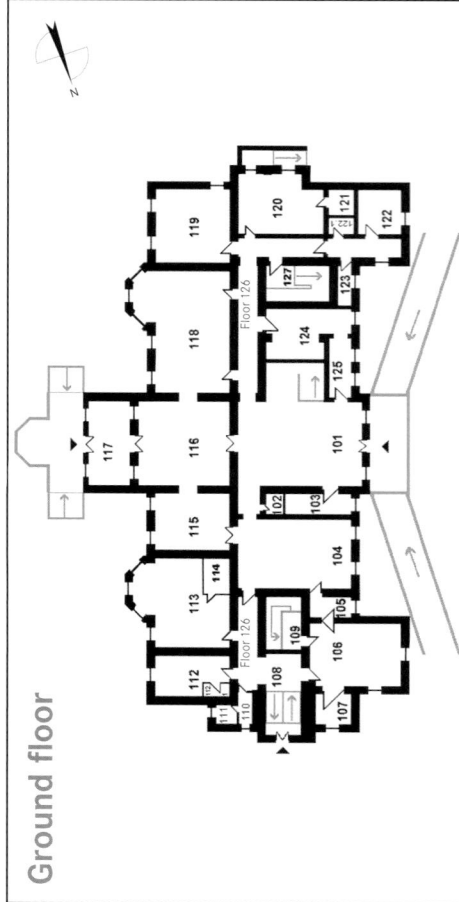

Basement

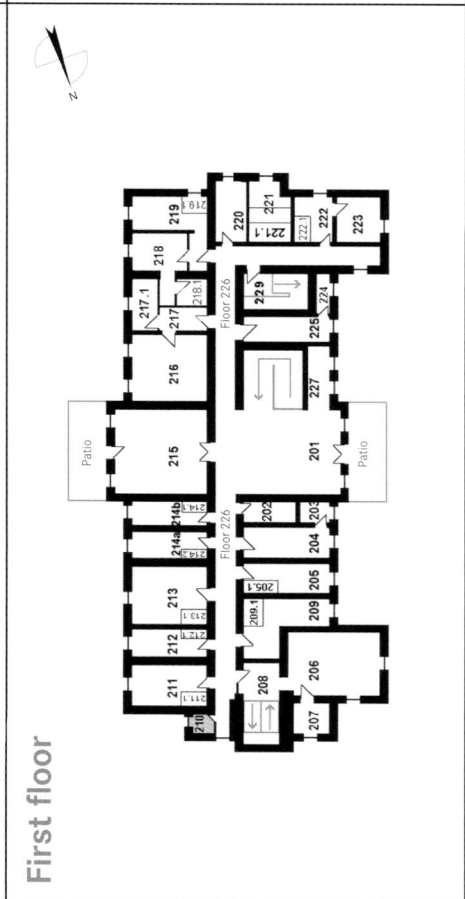

First floor

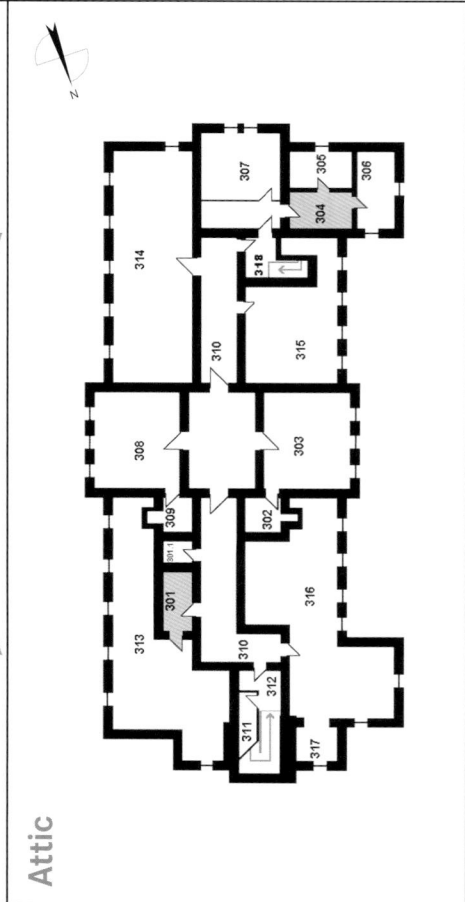

Routine cleaning

Frequency:

As required / min. every 10 days

First performed: May 05

Deep cleaning

Frequency: As required

First performed: May 05

Task	Work guidance	Additional comments	Material description / supply source
Routine cleaning of linoleum flooring	Sweep with broom and wipe down with slightly damp mop or cloth. Add standard proprietary cleaner to cleaning water.		
Deep cleaning of linoleum flooring	Standard proprietary LINO care product (coating material). Every 14 days adequate depending on duty.		
	Clean plastic skirting with mild scouring agent.		Material: 3.2 mm linoleum 3027.
			Supply source: enter company with address, tel/fax, e-mail and internet address.

Work location	**First floor: space 210** **Attic: spaces 301,304**		
Cross-references			
Issued on: 02.03.2007 Name: Miethe, Plätzer	**1st update on: 17.10.2007** Changes: Name: Frisch	**2nd update on: 17.05.2008** Changes: Name: Klemisch	**3rd update on:** Changes: Name:

Sheet no. 1/14

Ground floor

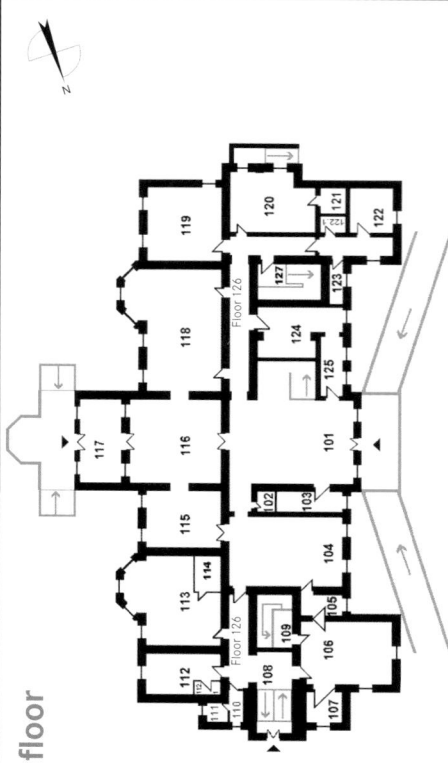

Attic

Basement

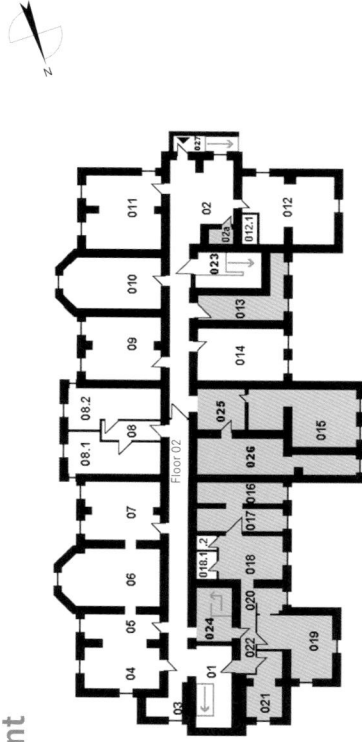

First floor

	Routine cleaning	Deep cleaning
Frequency:	As required / min. every 14 days	As required
First performed:	May 05	May 05

Task	Work guidance	Additional comments	Material description / supply source
Routine cleaning of screed	Sweep with broom.		
Deep cleaning of screed	Sweep with broom and wipe down with slightly damp mop or cloth. Add standard proprietary cleaner to cleaning water. Approx. once a month depending on duty and use.		

Work location	**Basement: spaces 02a, 013, 015, 016, 017, 018, 019, 020, 021, 022, 024, 025, 026**	

Cross-references		
Issued on: 02.03.2007 Name: Frisch	**1st update on: 15.07.2007** Changes: Name: Klemisch	**2nd update on:** Changes: Name:
		3rd update on: Changes: Name:

Ground floor

Attic

Basement

First floor

Routine cleaning		Deep cleaning	
Frequency:	As required / min. every 3 days	Frequency:	As required / every 14 days
First performed:	May 05	First performed:	May 05

Task	Work guidance	Additional comments	Material description/supply source
Routine cleaning of stairs	Dust regularly. Wipe down with damp (not wet!) cloth to remove dirt and then rub dry. Use only soft cloth.	Do not use ammonia-containing cleaners and solvents.	Use of mild cleaners, without scouring additives.
Protective treatment	After thorough cleaning, apply undiluted thin coat of gloss emulsion with soft cloth. Drying time approx. 20 minutes. Subsequent application intervals and method depending on product type; apply to cleaning water in diluted form, approx. 30 ml per 10 ltr, every 4–6 weeks for normal duty.	Under no circumstances should aggressive cleaners be used! Clean freshly coated (varnished) stairs only with mains water, with small quantities of mild cleaner (e. g. soft soap solution) added.	Gloss emulsion: enter make, address, tel/fax, e-mail and internet address.
Floor sealing	For stairs, every 8 years.	Subject to prior consultation with owner. To be performed only by suitable specialist contractor.	

Work location	Wooden stairs		
Cross-references			
Issued on: 02.03.2007	1st update on: 10.10.2007	2nd update on: 15.07.2008	3rd update on:
Name: Miethe, Pläzter	Changes: Name: Frisch	Changes: Name: Klemisch	Changes: Name:

Tower area 01

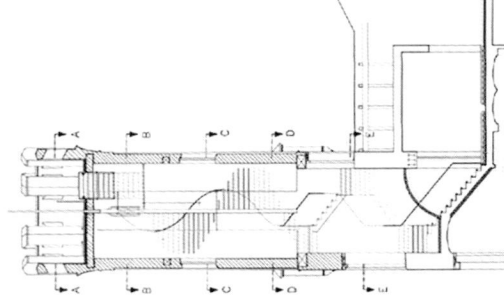

Basement staff stairs 024

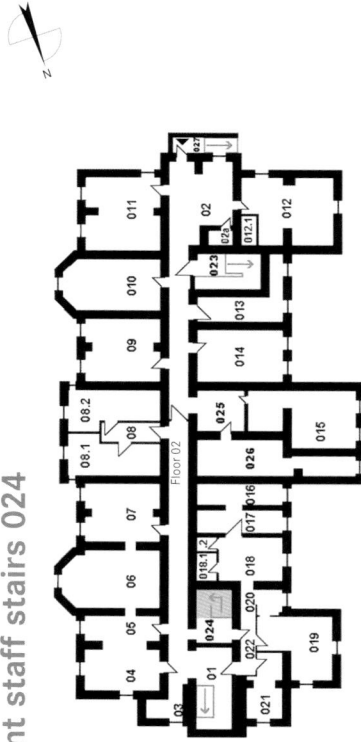

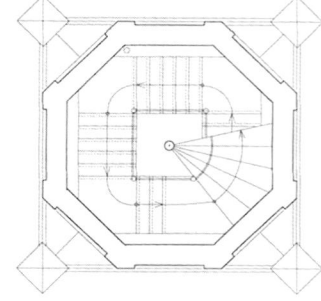

Section D - D

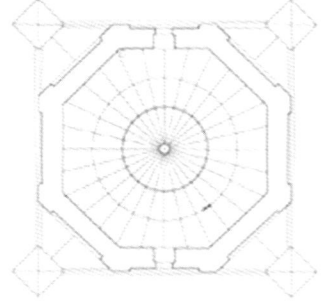

Section C - C

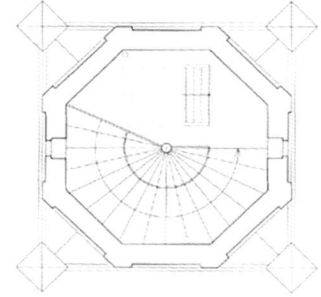

Section B - B

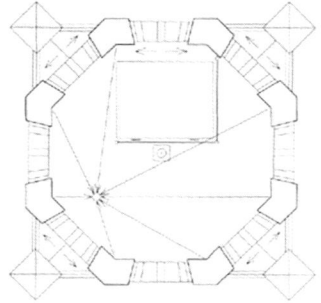

Section A - A

Routine cleaning

Frequency:

First performed: May 05

Deep cleaning

Frequency:

First performed: As required / every 14 days

May 05

Task	Work guidance	Additional comments	Material description / supply source
Routine cleaning of stairs	Sweep regularly with broom, remove dirt with damp mop or cloth.		Add standard proprietary cleaner to cleaning water.

Steps and landings in chequerplate, structural members and balustrade in hot-dip galvanized steel. |

| **Work location** | Basement staff stair – 024
Spiral tower stair | | |
|---|---|---|---|

Cross-references			
Issued on: 02.03.2007	1st update on: 15.07.2008	2nd update on:	3rd update on:
	Changes:	Changes:	Changes:
Name: Frisch	Name: Klemisch	Name:	Name:

Eastern façade

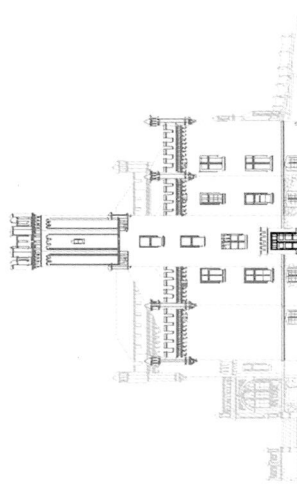

Eastern elevation

Western façade

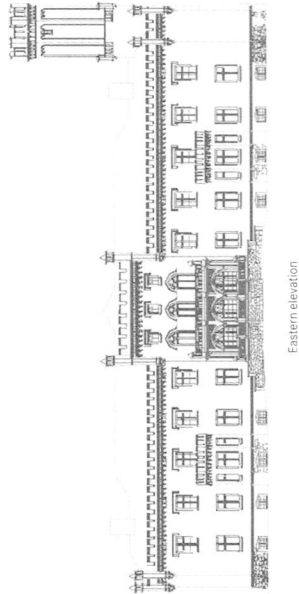

Western elevation

Northern façade

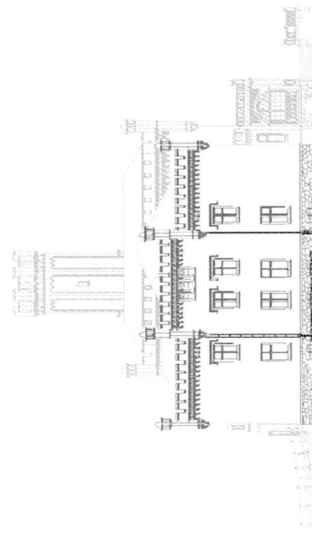

Southern façade

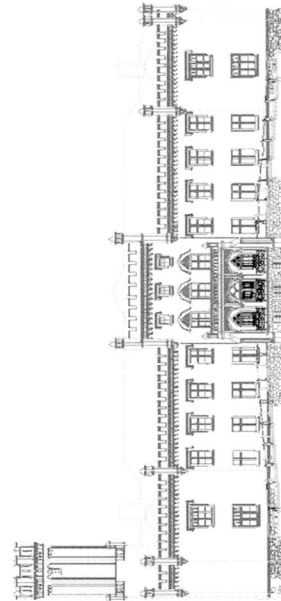

Routine cleaning

Frequency:	Monthly
First performed:	May 05

Stripping and repainting

Frequency	As required / every 14 days
First performed:	May 05

Task	Work guidance	Additional comments	Material description / supply source
Servicing of operating mechanisms	Regularly check for smooth operation and wear, grease/oil (once a year), check screws.		Acid- and resin-free grease.
Inspection of weatherstripping	Check at least every 6 months for brittle weatherstripping and mastic/putty joints.		
Inspection of weatherboards	Visually inspect condition of weatherboard coating annually.		
Windows, glazing	Perform regular visual inspection for any significant damage, also to pane (scratches, cracks).	Duty to notify owner. Call in specialist contractor where necessary.	
Cleaning of window panes (routine cleaning)	Clean with standard proprietary window-cleaning agents, as required by level of dirt.	Diluted with water in product-specific proportion.	
Cleaning of window frames (routine cleaning)	Clean with slightly damp cloth and standard proprietary acid-free cleaners at same time as window panes (not too damp).	Do not use any abrasive agents.	
Maintenance painting (overpainting)	Weatherboards every 2 years; on weathered elevations every 2 years, otherwise every 3 years. Scuff sand base in advance. Visually inspect for discoloration approx. every 6 months.	Subject to prior consultation with owner. To be performed only by suitable specialist contractor. Take out maintenance contract where appropriate. Do not overcoat weatherstripping.	Enter maintenance contractor. High-build alkyd woodstain, colour 048, rosewood. Basement storm windows, colour white, RAL 9010, silk; enter supply sources.
Stripping and repainting	To be performed approx. every 5–7 years following visual inspection (specialist contractor) where paintwork is found to be flaking. Perform hot-air stripping to remove existing paintwork (several coats) and soiling from entire surface. Burning-off with open flame is not permitted. Apply new coats of paint.	Subject to prior consultation with owner. To be performed only by suitable specialist contractor.	Priming coat with blue stain protection, alkyd woodstain first undercoat, apply surface filler, alkyd woodstain second undercoat, alkyd woodstain finishing coat. Make: high-build woodstain, colour 048, rosewood or white RAL 9010 (basement).

Work location	Timber windows		
Cross-references			
Issued on: 02.03.2007	**1st update on: 10.10.2007**	**2nd update on: 15.07.2008**	**3rd update on:**
Name: Miethe, Pläzer	Changes: Name: Frisch	Changes: Name: Klemisch	Changes: Name:

	Routine cleaning	Stripping and repainting	
Frequency:	Monthly		As required/indoor every 7 years, outdoor every 5 years
First performed:	May 05		May 05
Task	**Work guidance**	**Additional comments**	**Material description/supply source**
Servicing of operating mechanisms	Regularly check for wear, grease/oil, check screws.		Acid- and resin-free grease.
Inspection of weatherstripping and paintwork	Check at least every 6 months for brittle weatherstripping and mastic/putty joints.		
Windows, glazing	Perform regular visual inspection for any significant damage, also to pane (scratches, cracks).	Duty to notify owner. Call in specialist contractor where necessary.	
Cleaning of window panes (routine cleaning)	Clean with standard proprietary window-cleaning agents, as required by level of dirt.		Diluted with water in product-specific proportion.
Cleaning of window frames (routine cleaning)	Clean with slightly damp cloth and standard proprietary acid-free cleaners at same time as window panes (not too damp).		
Corrosion protection	Where paintwork is damaged, otherwise every 5 years. Scuff sand base, apply one coat of corrosion inhibitor, one priming coat, one undercoat and one finishing coat of alkyd paint.	Subject to prior consultation with owner. To be performed only by suitable specialist contractor.	1 x Redox AK 1190 (white). Rubbol primer (white). 1 x Color silk paint NCS-S 6010-G10Y. Supply source: enter specialist contractor.
Maintenance painting (overpainting)	Every 7 years (indoors) or 5 years (outdoors), over full surface (as specified above).	Subject to prior consultation with owner. To be performed only by suitable specialist contractor. Take out maintenance contract where appropriate.	Enter maintenance contractor with address, tel/fax, e-mail and internet address.

Work location	Steel windows in basement		
Cross-references			
Issued on: 02.03.2007	**1st update on:** 10.10.2007	**2nd update on:** 15.07.2008	**3rd update on:**
	Changes:	Changes:	Changes:
Name: Miethe, Plätzer	Name: Frisch	Name: Klemisch	Name:

Routine cleaning

Frequency: Once a year

First performed: May 05

Stripping and repainting

Frequency: Every 5–10 years

First performed: May 05

Task	Work guidance	Additional comments	Material description / supply source
Inspection of paintwork	Perform regular visual inspection for rust and flaking paint, at least once a year.		
Cleaning	Brush down grilles twice a year.		
Corrosion protection	Where paintwork is damaged, otherwise every 5 years. Scuff sand base, apply one coat of corrosion inhibitor, one priming coat, one undercoat and one finishing coat of alkyd paint.	Subject to prior consultation with owner. To be performed only by suitable specialist contractor.	1 x Sikkens Redox AK 1190 (white). 2 x Sikkens Rubbol primer (white). 1 x Sikkens Color silk paint NCS-S 6010-G10Y.
Maintenance painting (overpainting)	Every 10 years, over full surface (as described above).	Subject to prior consultation with owner. To be performed only by suitable specialist contractor. Take out maintenance contract where appropriate.	Supply source: enter address, tel / fax, e-mail and internet address.
Work location	External windows		

Cross-references			
Issued on: 02.03.2007	1st update on: 10.10.2007	2nd update on: 15.07.2008	3rd update on:
Name: Miethe, Pläzter	Changes: Name: Frisch	Changes: Name: Klemisch	Changes: Name:

Ground floor

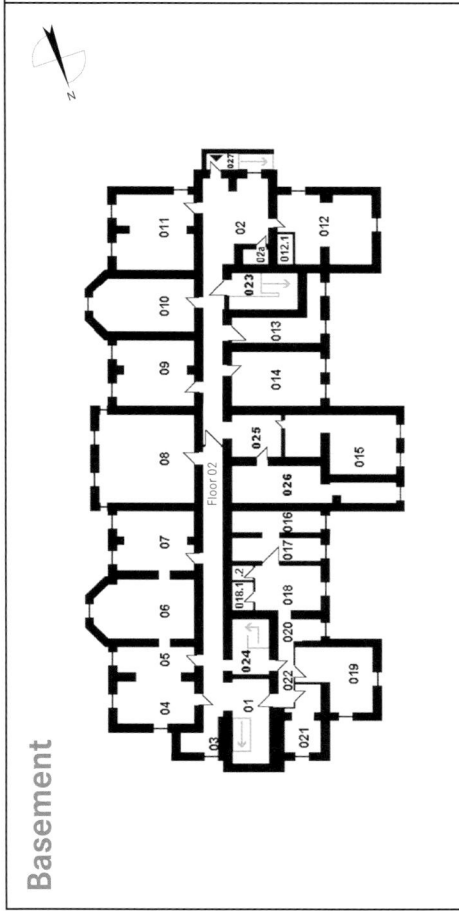

Attic

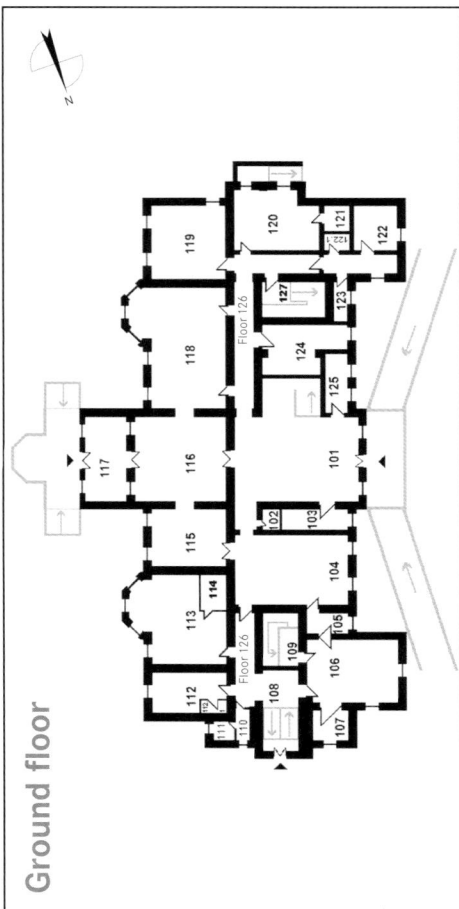

Basement

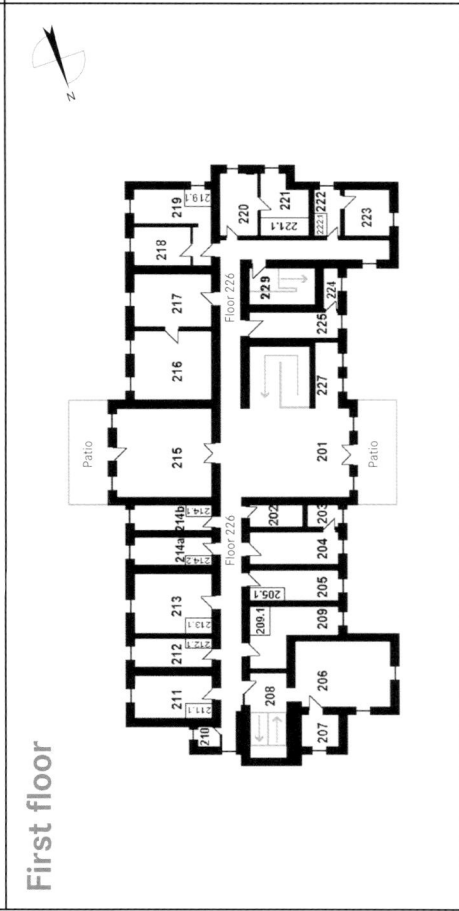

First floor

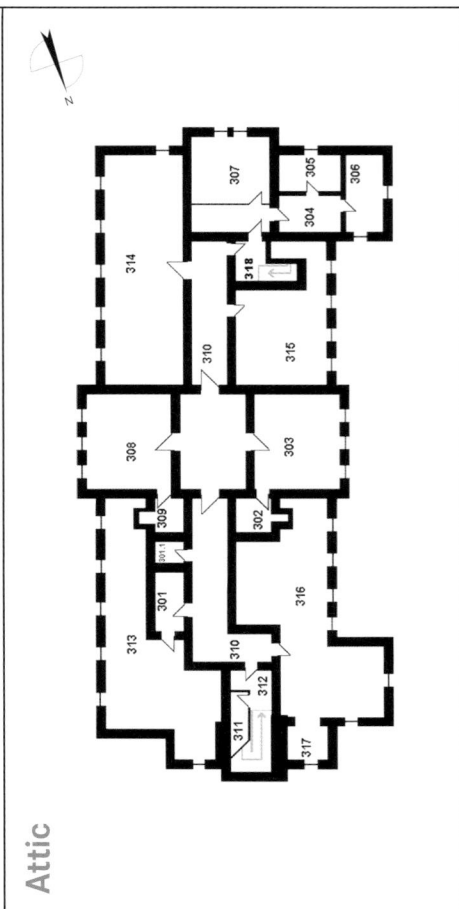

Routine cleaning		Stripping and repainting		
Frequency:	Monthly	Frequency:		Indoor every 7 years / outdoor min. every 3 years
First performed:	May 05	First performed:		May 05

Task	Work guidance	Additional comments	Material description / supply source
Servicing of operating mechanisms	Regularly check for wear, grease / oil (once a year), check screws.		Panelled door with solid pine panels, double-leaf with active and inactive leaf, fixed fanlight.
Servicing of door closers of fire and escape doors	Adjust door closers at least once a year.	To be performed by specialist contractor. Take out maintenance contract.	
Inspection of weatherstripping	Check at least every 6 months for brittle weatherstripping and mastic / putty joints.		
Inspection of external door weatherboards	Visually inspect condition of weatherboard coating, annually.		
Doors, glazing	Perform regular visual inspection for any significant damage, also to glazing (scratches, cracks).	Duty to notify owner. Call in specialist contractor where necessary.	
Cleaning of door surfaces	Clean with standard proprietary acid-free cleaners, diluted with water, as required by degree of soiling.	Do not use any abrasive agents.	
Maintenance painting (overpainting)	Weatherboards every 2 years; on weathered elevations every 3 years; internal face every 7 years. Scuff sand base in advance. Visually inspect for discoloration approx. every 6 months for external doors.	Maximum 3 x maintenance coats, then stripping and repainting in accordance with VOB/C. Painting by specialist contractor. Consultation with owner required.	Contractor: enter address, tel/fax, e-mail and internet address.Coating: high-build woodstain, colour 048, rosewood. Enter supply source.
Stripping and repainting	To be performed following visual inspection (specialist contractor) where paintwork is found to be flaking. Perform hot-air stripping to remove existing paintwork (several coats) and soiling from entire surface. Burning-off with open flame is not permitted. Apply new coats of paint. Surface levelness to DIN standard, alkyd woodstain.	Subject to prior consultation with owner. To be performed only by suitable specialist contractor.	Priming coat with blue stain protection (outdoors), otherwise alkyd woodstain (indoors), first alkyd woodstain undercoat (indoors and outdoors), second alkyd woodstain undercoat (outdoors), alkyd woodstain finishing coat (indoors and outdoors), high-build woodstain, colour 048, rosewood. Internal door ironmongery: Montparnasse model. Enter supply source. Original lever handles: enter company/designation.

Work location	Internal / external doors		
Cross-references	Timber windows, sheet no. 3/2		
Issued on: 02.03.2007	**1st update on: 10.10.2007**	**2nd update on: 15.07.2008**	**3rd update on:**
	Changes:	Changes:	Changes:
Name: Miethe, Plätzer	Name: Frisch	Name: Klemisch	Name:

Sheet no. 4/2

Ground floor

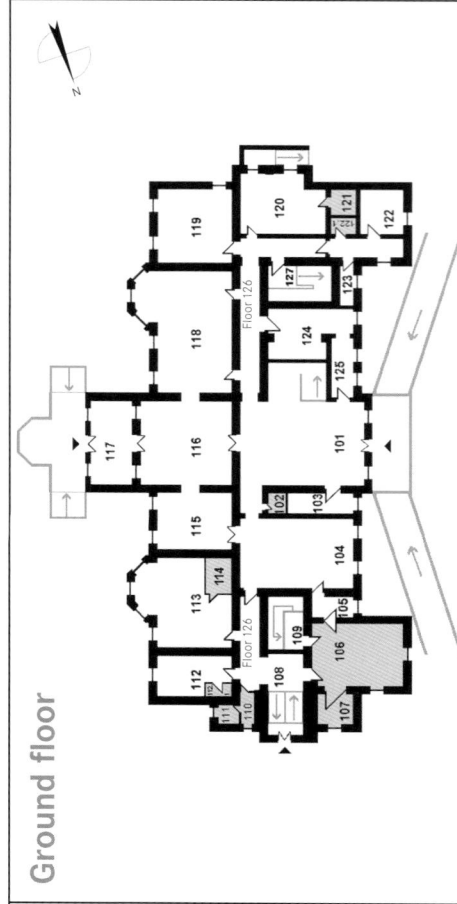

Attic

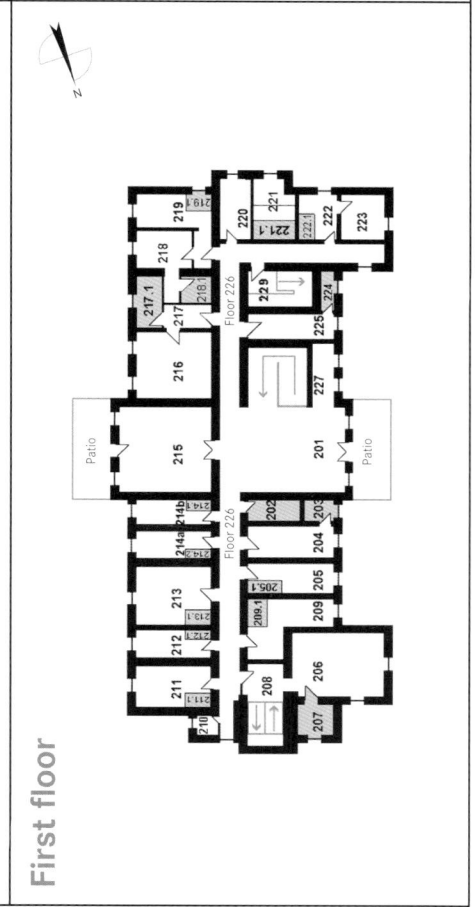

Basement

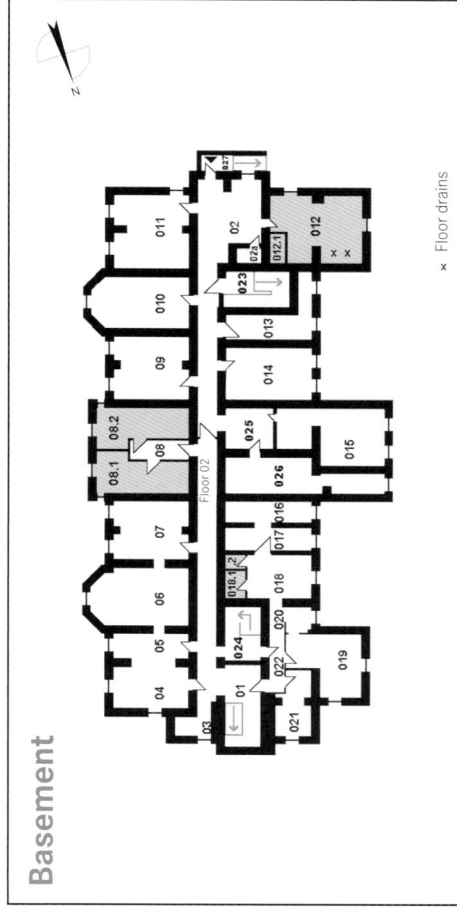

× Floor drains

First floor

Routine cleaning

Frequency: As required / min. every 2 days

First performed: May 05

Deep cleaning

Frequency: As required / min. every 14 days

First performed: May 05

Task	Work guidance	Additional comments	Material description / supply source
Inspection	Visually inspect joint seals in conjunction with cleaning.	Cleaning staff, caretaker.	Standard proprietary tile cleaner and descaler.
Routine / deep cleaning of tiling	Sweep with broom and wipe down with slightly damp mop or cloth. Add tile cleaner to cleaning water. Clean tiles with descaling agent every 14 days.		
Cleaning of floor drains (see x in floor plan)	Remove dirt from strainers every day when room is occupied. Rinse floor drains with water every month to prevent build-up of odours.		
Renewal of joint seals	To be performed when peeling and cracks are visible. In areas exposed to splashing, approx. every 2–4 years, otherwise approx. every 4–10 years.	To be performed by suitable specialist contractor. Maintenance contract with specialist contractor recommended, for periodic (half-yearly) renewal of damaged areas only.	Silicone sealant. Tiling: Enter company address, tel / fax, e-mail and internet address; size 200 x 200 mm. Enter details of any maintenance contractor.
Work location	**Basement: spaces 08.1, 08.2, 012, 012.1, 018.1, 018.2** **Ground floor: spaces 102, 110, 111, 112.1, 114, 121, 122.1 and kitchen 106, 107 (wall tiles)** **First floor: spaces 202, 203, 205.1, 207, 209.1, 211.1–214.2, 217.1–219.1, 221.1, 222.1, 224; Attic: spaces 301.1, 302, 305, 309**		
Cross-references	Sanitary appliances, sheet no. 6 / 1–6 / 2; Floors / tiling, sheet no. 1 / 9–1 / 10		
Issued on: 02.03.2007 Name: Miethe, Plätzer	**1st update on: 17.10.2007** Changes: Name: Frisch	**2nd update on: 15.07.2008** Changes: Name: Klemisch	**3rd update on:** Changes: Name:

Work card

325 Sanitary appliances
Sheet no. 6 / 1

Basement

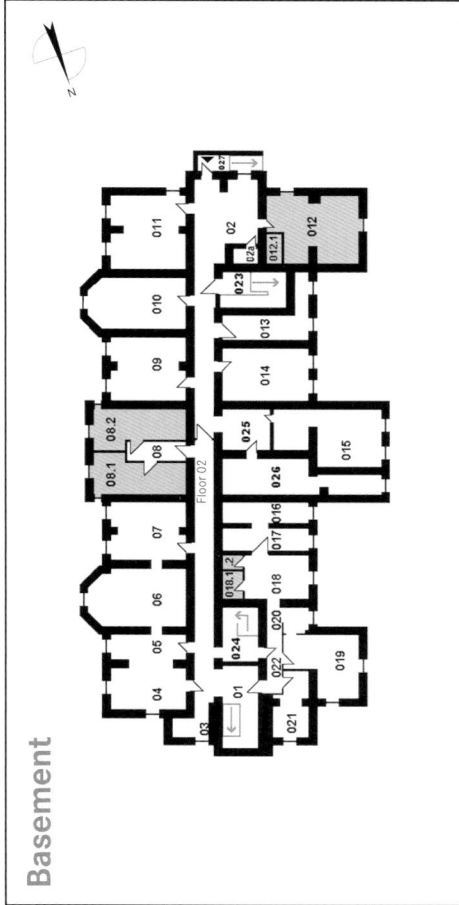

Ground floor

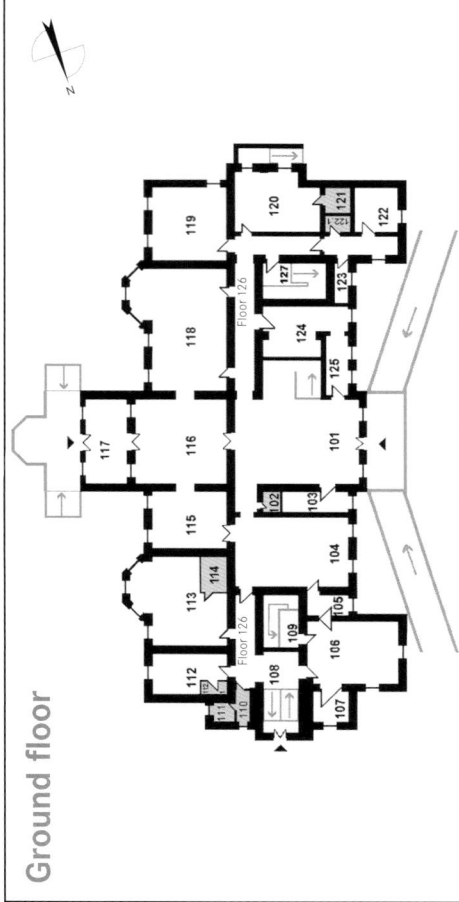

First floor

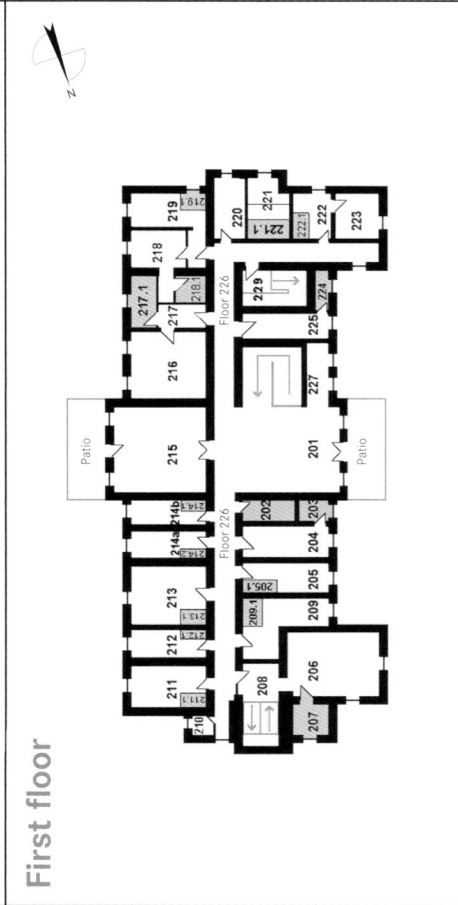

Attic

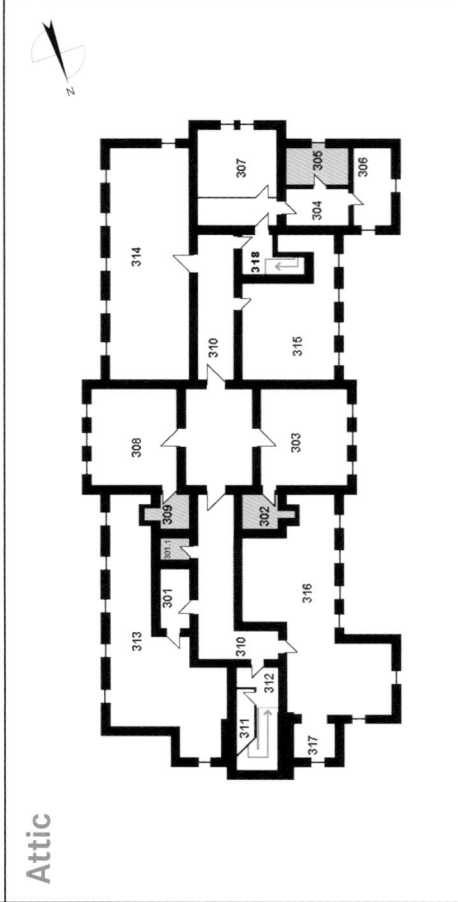

Inspection		Routine cleaning	
Frequency:		Frequency:	
As required / weekly		As required / min. every 2 days; 2–5 years	
First performed:		First performed:	
May 05		May 05	

Task	Work guidance	Additional comments	Material description / supply source
Inspection	Visually inspect for leaks, in conjunction with cleaning. 1 x annual inspection – and, where necessary, cleaning of – wash basin, shower and bathtub traps, seals to water fittings, grab rails, toilet support rails etc. (renewal approx. every 2–5 years).	Cleaning staff, caretaker. Maintenance contract recommended for tiling seals.	Enter details of any maintenance contractor.
Silicone joint seals to showers, bathtubs	Check silicone joint seals for proper adhesion by visual and tactile inspection during routine cleaning. Renew any damaged seals.	Cleaning staff, caretaker, maintenance contract recommended, as above.	Silicone sealant.
Routine cleaning of sanitary appliances	Clean with ceramic-compatible household cleaner or mild soapy water using soft cloth. Clean sanitary appliances with gentle descaler every 14 days.	Do not use abrasive or corrosive cleaners.	Standard cleaners for sanitary appliances: Villeroy & Boch "Basic" model (wall-hung WC pan, washbasin, bathtub), "Evana" model (undercounter washbasin), "Ebra" model (urinal), "Cosmos" model (shower trays).
Routine cleaning of toilet seat, flushing cistern	Clean with plastics-compatible household cleaner or mild soapy water using soft cloth. Do not allow toilet cleaners to come into contact with hinges or other fittings. When toilet cleaner is used, lift up toilet lid and seat and do not put back down until toilet cleaner has been completely flushed down WC pan.	Do not use abrasive or corrosive cleaners.	Standard cleaners.
Routine cleaning of water fittings	Clean with soapy water using soft cloth or leather. Gently remove heavier soiling or scale with household vinegar or sanitary cleaner. Apply to surface, leave to act for a short time, then thoroughly rinse with clean water.	Under no circumstances should cleaners containing alcohol or gritty abrasive be used. Soft, non-abrasive sponges should also be used.	Standard cleaners for water fittings: Single-lever mixer taps: enter make, address, tel / fax, e-mail and internet address.
Shower hair traps	Clean "from top" on daily basis or upon change of room occupancy.	Cleaning staff.	

Work location			
Basement: spaces 08.1, 08.2, 012, 012.1, 018.1, 018.2 **Ground floor: spaces 102, 110, 111, 112.1, 114, 121, 122.1** **First floor: spaces 202, 203, 205.1, 207, 209.1, 211.1–214.2, 217.1–219.1, 221.1, 222.1, 224** **Attic: spaces 301.1, 302, 305, 309**			

Cross-references			
Tiling to sanitary areas, sheet no. 5 / 1–5 / 2			

Issued on: 02.03.2007	1st update on: 15.07.2008	2nd update on:	3rd update on:
	Changes:	Changes:	Changes:
Name: Frisch	Name: Klemisch	Name:	Name:

Ground floor

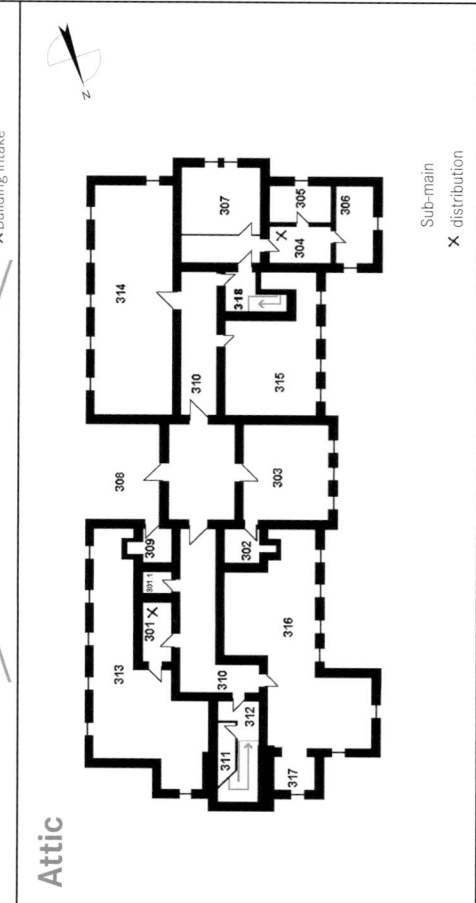

X Building intake

Attic

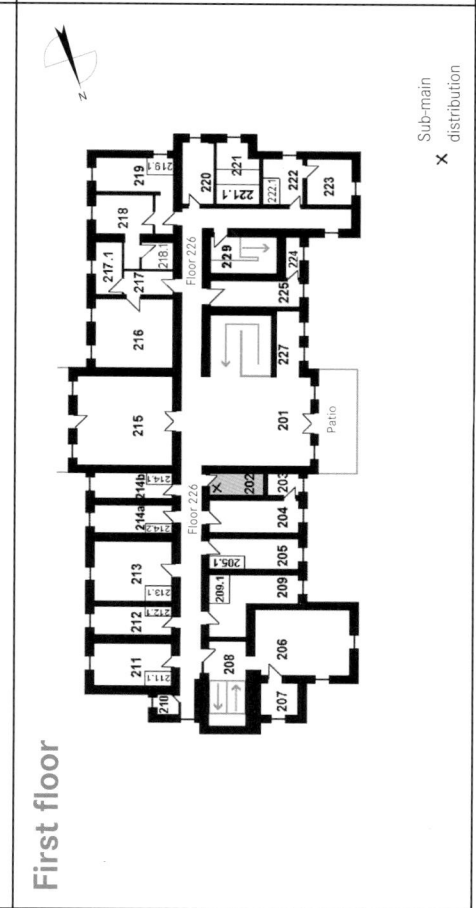

Sub-main
X distribution

Basement

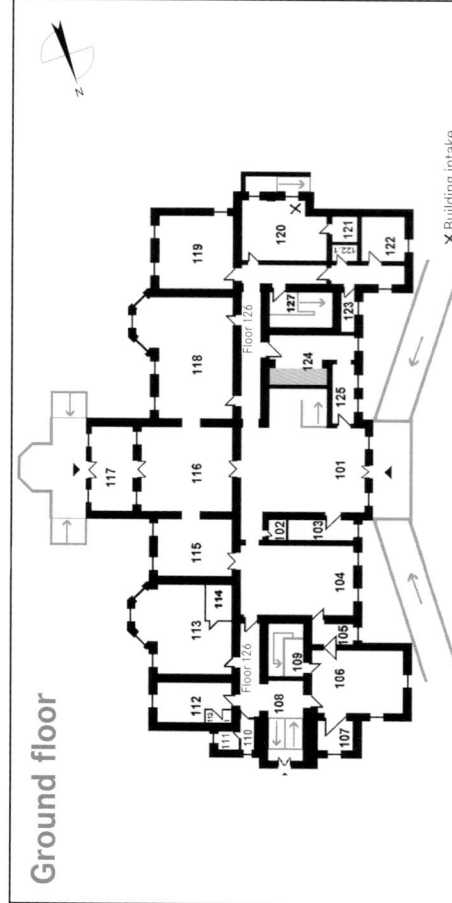

Switch cabinet –
remote control panel

Electrical switch cabinet for
sewage treatment system

Vebtilation switch
cabinet

Kitchen

sub-distribution board

First floor

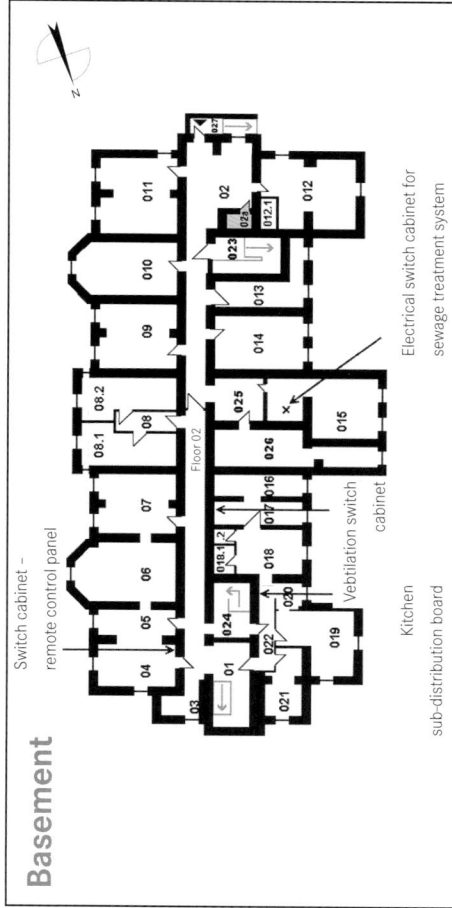

Sub-main
X distribution

Electrical installations

Inspection		Servicing	
Frequency: See work guidance		Frequency: See work guidance	
First performed: May 05		First performed: May 05	

Task	See work guidance	Additional comments	Material description / supply source
Inspection	Visually inspect for damage to switches and socket outlets. Check for fit condition and damage to insulation. Inspection in generally accessible areas. Once a month.	Call in specialist contractor in case of defects.	
Electrical installation	Check installations to DIN 015 Part 1 and DIN 108 as well as UVV VBG 4. Once a year.	Take out maintenance contract with specialist contractor.	Maintenance contractor: enter address, tel/fax, e-mail and internet address.
Residual-current circuit-breakers (rccb)	Perform operating test to DIN 108 Part 1 and VBG 4. Activate function test button, keep log. Once a year.	Take out maintenance contract with specialist contractor.	Maintenance contractor: enter address, tel/fax, e-mail and internet address.
Luminaires (indoor and outdoor)	Check fit condition, particularly regarding time switches, replace luminaires or lamps where necessary, check for adequate brightness. Once a month.	By caretaker	See schedule of installed luminaires.
Switch and socket outlet range	Permanently monitor for any faults. Check on fault panel during inspection of electrical installation.		Make: standard, electric white. Enter supply source.
Fault panel		Take out maintenance contract with specialist contractor.	Maintenance contractor: enter address, tel/fax, e-mail and internet address.
Lightning protection	Take measurements at test points / ground entries, every 2 years.	Take out maintenance contract w	Maintenance contractor: enter address, tel/fax, e-mail and internet address.
Emergency lighting, escape route symbols	Check for proper functioning, visually inspect, keep log. Weekly inspection.	By caretaker.	
Fire alarm system	Check system to DIN 015 Part 1 and DIN 108, 4 times a year.	Take out maintenance contract with specialist contractor.	Maintenance contractor: enter address, tel/fax, e-mail and internet address.
Fire extinguishers	Replace every 2 years.	Take out maintenance contract with specialist contractor.	Enter supply source, maintenance contractor: dry powder fire extinguisher PG 6 l/carbon-dioxide fire extinguisher.
Work location	Electrical plantroom: in basement 02a; in ground floor space 124: electrical control station, telecom and fire-alarm distribution boards Sub-main distribution: space 202 (1F), vestibule to 219 (1F), spaces 304, 301 (A)		
Cross-references			

Issued on: 02.03.2007	1st update on: 15.07.2008	2nd update on:	3rd update on:
	Changes:	Changes:	Changes:
Name: Miethe, Pläzter	Name: Klemisch	Name:	Name:

325/352

Ground floor

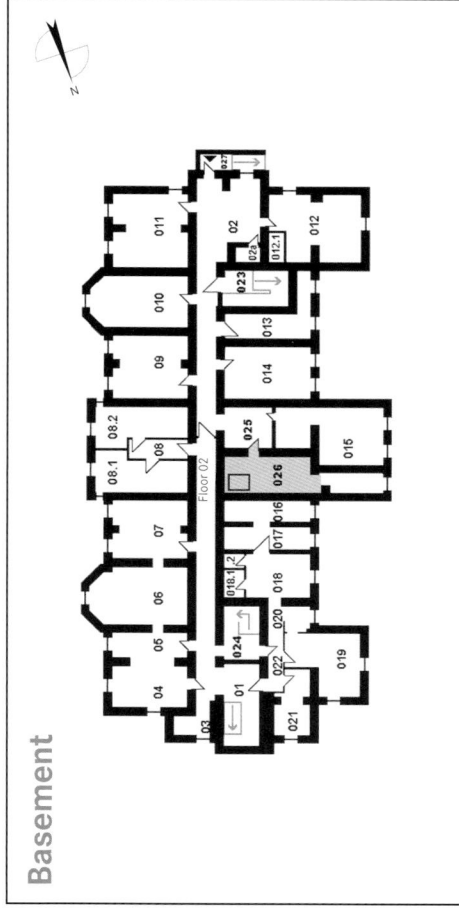

Attic

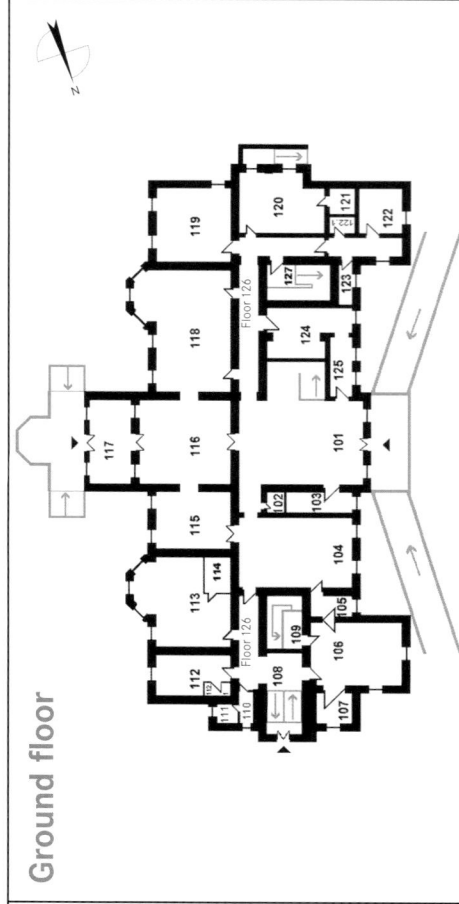

Basement

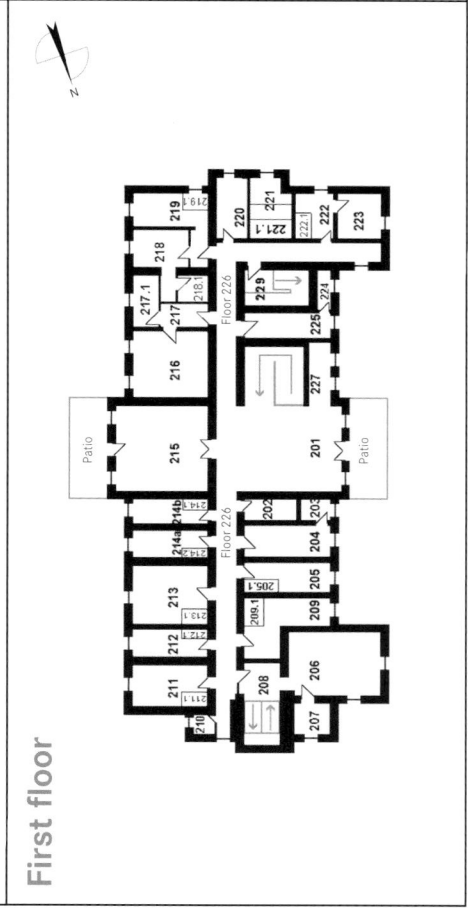

First floor

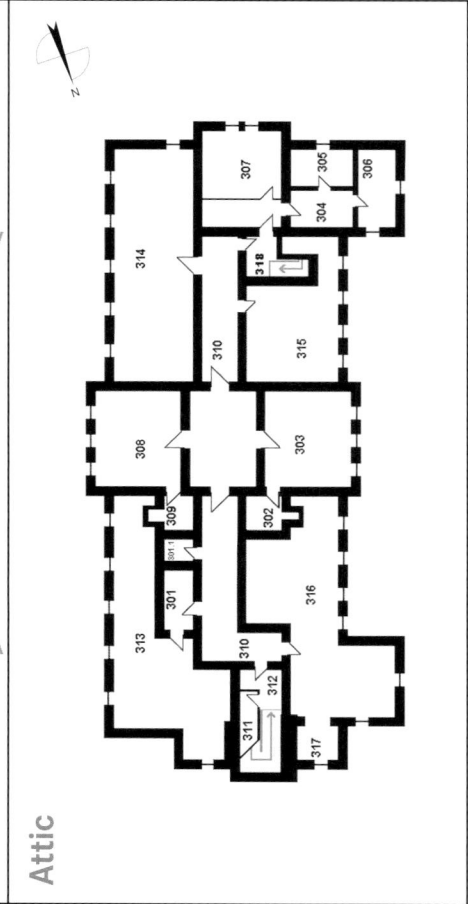

Routine cleaning	Maintenance painting		
Frequency: Every 14 days	Frequency:		As required
First performed: May 05	First performed:		May 05

Task	Work guidance	Additional comments	Material description / supply source
Servicing of burner	Twice a year.	Take out maintenance contract. To be performed by specialist contractor.	Burner: Burner type VT 2 a II – G. Enter maintenance contractor.
Pressure check, boiler, storage cylinder	Inspect once a month; perform thermal sterilization once a year.	Take out maintenance contract. To be performed by specialist contractor.	Boiler: Eurobloc EB 240 / Eurotronic D control system. Hot-water cylinder. 2 EAS cylinders, 300 to 500 ltr.
System inspection	Check for noises, open bleed valves, add top-up water to maintain pressure; check thermostatic valves once a year for proper functioning. Check of valves and connections, twice a year. Visually inspect for missing insulation, loose supports, once a month.	Call in specialist contractor or specially trained personnel. For heating system servicing instructions and inspection / servicing plan, see annex.	Manufacturer and maintenance contractor: enter address, tel / fax, e-mail and internet address.
Cleaning	Wipe down with slightly damp cloth, check for flaking paint and leaks, every 14 days.		
Maintenance painting and repainting of heat emitters and flow / return pipes	Maintenance painting: touch up localized areas of flaking paintwork. Repainting based on visual assessment: brush and roller coating of all surfaces.	Consultation with owner required. To be performed by specialist contractor. Take out maintenance contract.	Radiator enamel P 822, colour RAL 9010 or colour of walls. Heat emitter make: ARBONIA. Supply source: enter specialist contractor.
Heating flues	Inspect heating flues, once annually.	Visual inspection by chimney sweep.	Enter address, tel / fax, e-mail and internet address of chimneysweep.

Work location	Heating plantroom / boiler in basement space 026		
Cross-references			
Issued on: 02.03.2007	**1st update on: 11.10.2007** Changes: Name: Frisch	**2nd update on: 15.07.2008** Changes: Name: Klemisch	**3rd update on:** Changes: Name:
Name: Miethe, Plázer			

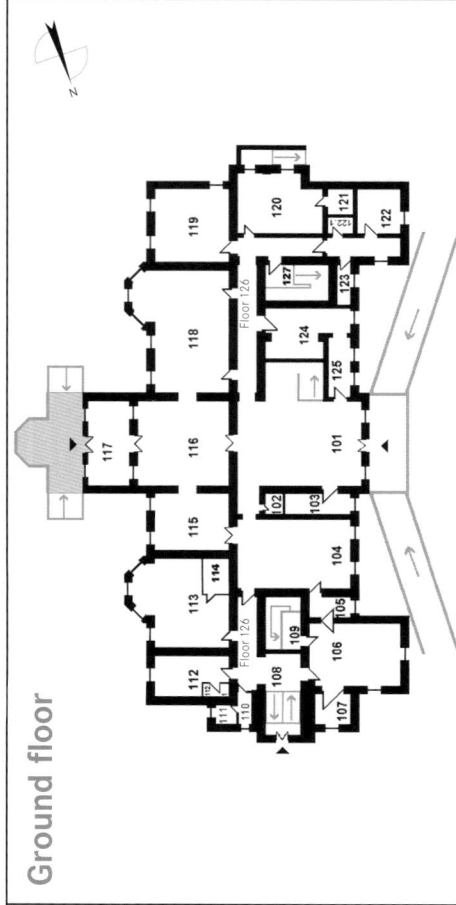

Ground floor

Attic

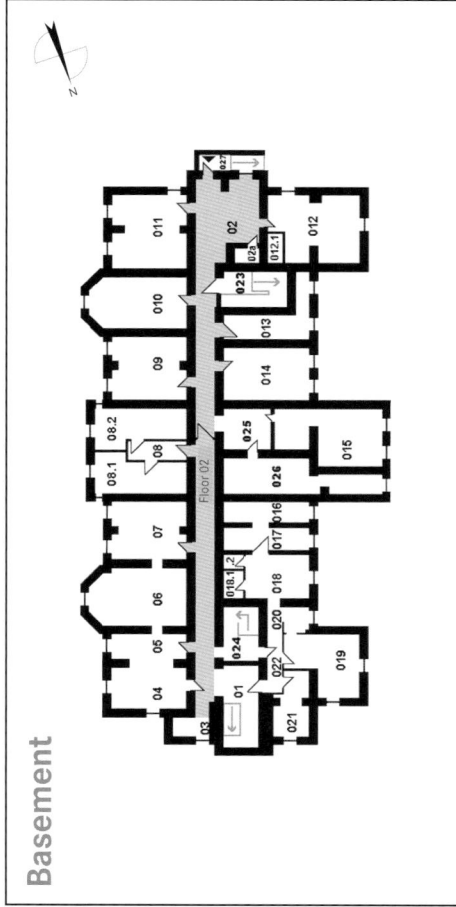

Basement

First floor

Special elements, miscellaneous

Inspection	Routine cleaning	
Frequency:	Frequency:	
First performed:	First performed:	
See work guidance	See work guidance	See work guidance
May 05	May 05	May 05

Task	See work guidance	Additional comments	Material description/supply source
Patios	Inspect downpipes, rainwater gutters and gullies, after heavy rainfall or once a month.	By caretaker. Duty to notify owner.	
	Visually inspect waterproofing and capping strips	By caretaker. Duty to notify owner.	
Coffered ceiling in basement	Open only for repair purposes.	By specialist contractor or specially trained personnel. Wear safety gloves.	Replacement coffers provided. 'Kontur D 147' coffer ceiling, smooth, 600 x 600 mm; enter supply source.
Inspection of basement/roof spaces for prevention of dry rot	Regular visits and visual inspection by operator; check for regular ventilation and adequate heating; in case of damage, repairs to be performed by specialist contractor.	Duty to notify owner in case of abnormalities. Call in suitable specialist contractor.	Waterproofing mortar: Aida repellent mortar (to seal open joints and voids). Enter details of any specialist contractor.
	Inspect exposed timbers, notify of any leaks/bursts and drying out measures, call in timber surveyor where necessary. Perform inspection with timber surveyor every 2 years.	Duty to notify owner in case of abnormalities. Call in suitable specialist contractor. Take out inspection/ maintenance contracts with dry rot surveyor.	Timber surveyor: enter address, tel/fax, e-mail and internet address.
Dry rot treatment	Call in specialist contractor.	Consultation with owner required.	Adolit M dry rot treatment. Enter supply source, specialist contractor's address, tel/fax, e-mail and internet address.

Work location	Patios, coffered ceiling in basement hallway 02		
Cross-references	External drainage, sheet no. 11/4		
Issued on: 02.03.2007	1st update on: 10.10.2007	2nd update on: 15.07.2008	3rd update on:
	Changes:	Changes:	Changes:
Name: Frisch	Name: Frisch	Name: Klemisch	Name:

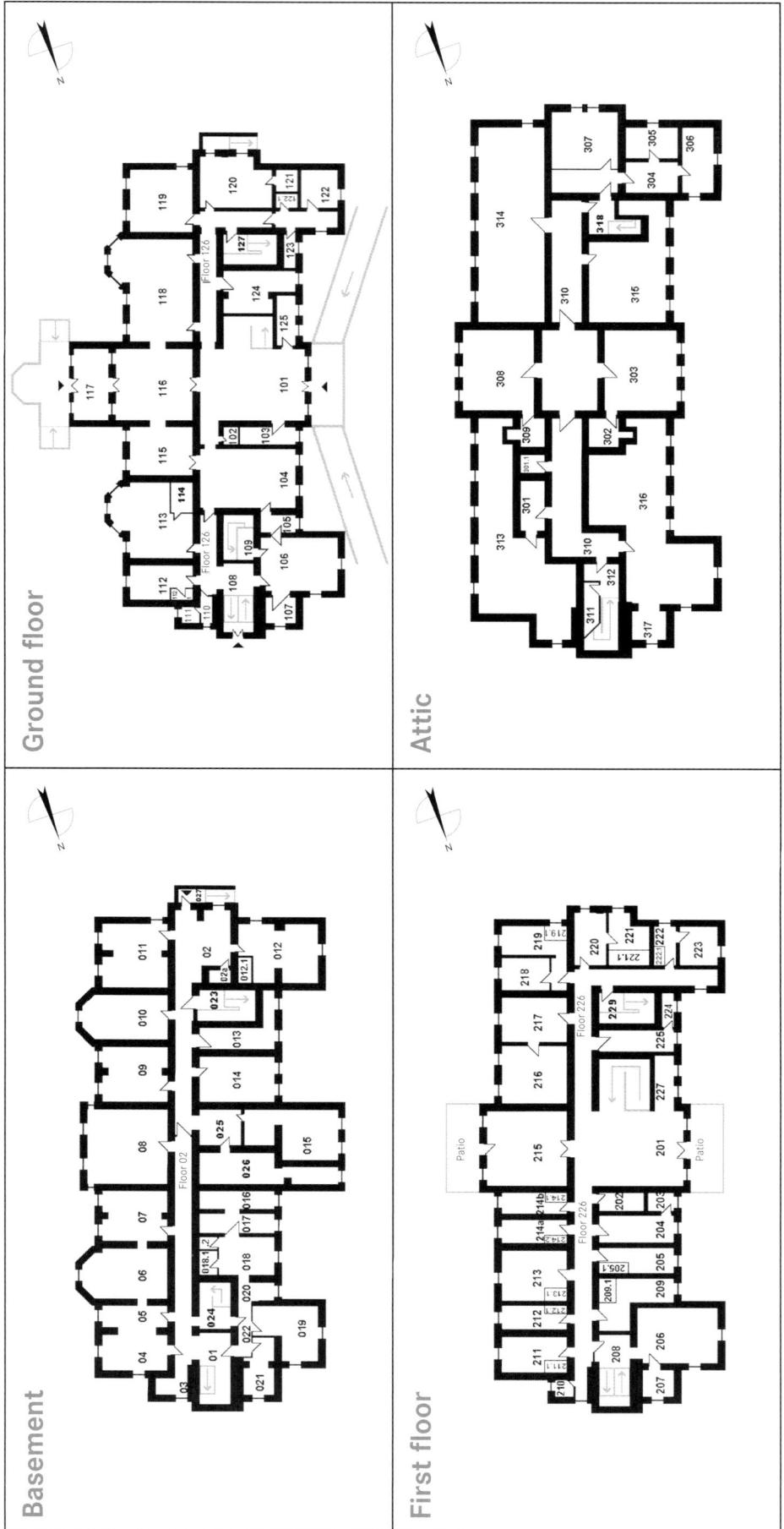

Ground floor

Attic

Basement

First floor

Inspection			Deep cleaning	
Frequency:			Frequency:	As required / min. every 14 days
Twice a year				
First performed:			First performed:	
May 05			May 05	

Task	Work guidance	Additional comments	Material description/supply source
Inspection	Visually inspect twice a year, check for operability and easy closing, check on wear.	By caretaker.	
Deep cleaning	Clean with dry duster as required or at least every 14 days.		Radiator cover: Timber radiator cover in window recesses with double-leaf, perforated metal doors, timber species: woodstained pine.

Work location		

Cross-references		
Issued on: 02.03.2007	1st update on: 15.07.2008	2nd update on:	3rd update on:
	Changes:	Changes:	Changes:
Name: Frisch	Name: Klemisch	Name:	Name:

Ground floor

Attic

Basement

× Floor drains

First floor

Inspection		Deep cleaning	
Frequency:	Twice a month	Frequency:	Four times a year
First performed:	May 05	First performed:	May 05

Task	Work guidance	Additional comments	Material description / supply source
Floor drains (see x in floor plan), traps	Pour in water, also in spaces unused for long time, four times a year.	By caretaker.	
Waste water traps	Check water seal in traps; rinse filter, every 3 weeks.		
Inspection	Check silicone joints for proper, watertight seal; check for proper run-off.	Maintenance contract recommended for tiling seals.	Enter details of any maintenance contractor, with address, tel/fax, e-mail and internet address
Cleaning	Call in specialist contractor to deal with any malfunctions, soiled pipes and damaged seals.		
Work location	**Bathrooms, kitchens** **B: spaces 08.1, 08.2, 012, 012.1, 018.1, 018.2** **GF: spaces 102, 110, 11, 112.1, 114, 121, 122.1 and kitchen 106, 107** **1F: spaces 202, 203, 205.1, 207, 209.1, 211.1–214.2, 217.1, 219.1, 221.1, 222.1, 224** **A: spaces 301.1, 302, 305, 309**		

Cross-references	Tiling to sanitary areas, sheet no. 5/1–5/2		
Issued on: 02.03.2007	1st update on: 17.10.2007	2nd update on: 15.07.2008	3rd update on:
	Changes:	Changes:	Changes:
Name: Miethe, Plätzer	Name: Frisch	Name: Klemisch	Name:

Eastern façade

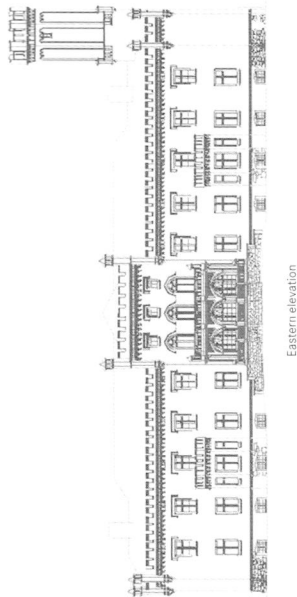

Eastern elevation

Western façade

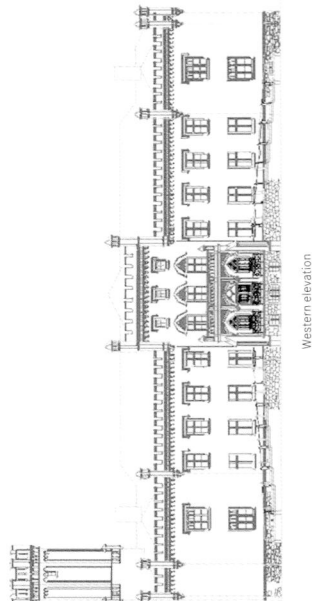

Western elevation

Northern façade

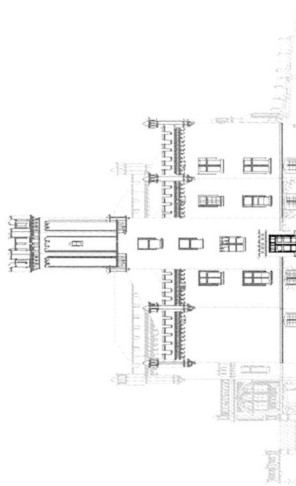

Southern façade

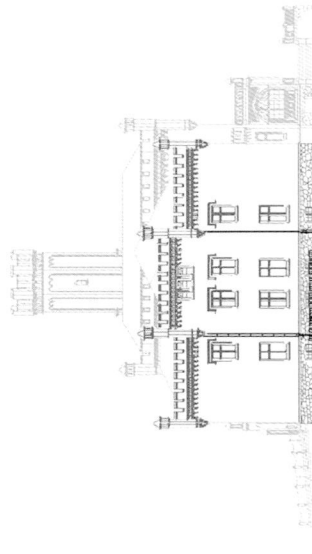

Routine cleaning		Servicing	
Frequency:	See work guidance	Frequency:	See work guidance
First performed:	May 05	First performed:	May 05
Task	**Work guidance**	**Additional comments**	**Material description / supply source**
Inspection / cleaning of gutters	Visually inspect through roof hatches and from tower for being watertight, fit condition and soiling, once a month. Immediately remove any clogging in or damage to system. Particularly after autumn leaf fall. Clean in mid-December and April.	By caretaker.	
Inspection of internal rainwater downpipes	Visually inspect downpipes in attic/loft space, after heavy rainfall or once a month.	By specialist contractor or specially trained personnel. Take out maintenance contract with specialist contractor where appropriate.	Enter details of any maintenance contractor, with address, tel/fax, e-mail and internet address.
Patio drainage	Inspect downpipes, gutters and gullies, after heavy rainfall or once a month. Visually check for integrity of waterproofing.	By caretaker. Duty to notify owner. By caretaker. Duty to notify owner.	Enter details of any maintenance contractor, with address, tel/fax, e-mail and internet address.
Cleaning of downpipes	Open cleaning eyes in base standpipe section of downpipes and remove soiling, as required or twice a year (November and March). Brush down external downpipe surfaces on façade, as required or once a year.	By specialist contractor or specially trained personnel. Take out maintenance contract with specialist contractor where appropriate. By caretaker or included in maintenance contract with specialist contractor.	
Rinsing of perimeter drains	Visually inspect, open, rinse, close, once a year.	Take out maintenance contract with specialist contractor.	
Standpipes	Inspect and clean standpipes (at base of downpipes) four times a year.	By specialist contractor or specially trained personnel. Take out maintenance contract with specialist contractor where appropriate.	Enter details of any maintenance contractor, with address, tel/fax, e-mail and internet address.
Light wells	Check for proper run-off of rainwater, after heavy rainfall or once a month.	By caretaker. Duty to notify owner.	
Work location	Façade, eaves area behind crenellations, plinth area, light wells		
Cross-references			
Issued on: 02.03.2007	**1st update on: 11.10.2007**	**2nd update on: 15.07.2008**	**3rd update on:**
Name: Miethe, Pläzter	Changes: Name: Frisch	Changes: Name: Klemisch	Changes: Name:

Ground floor

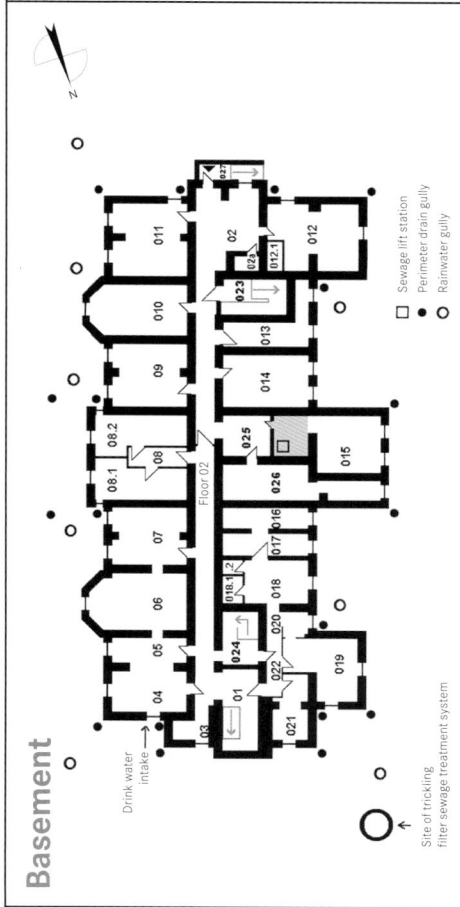

Attic

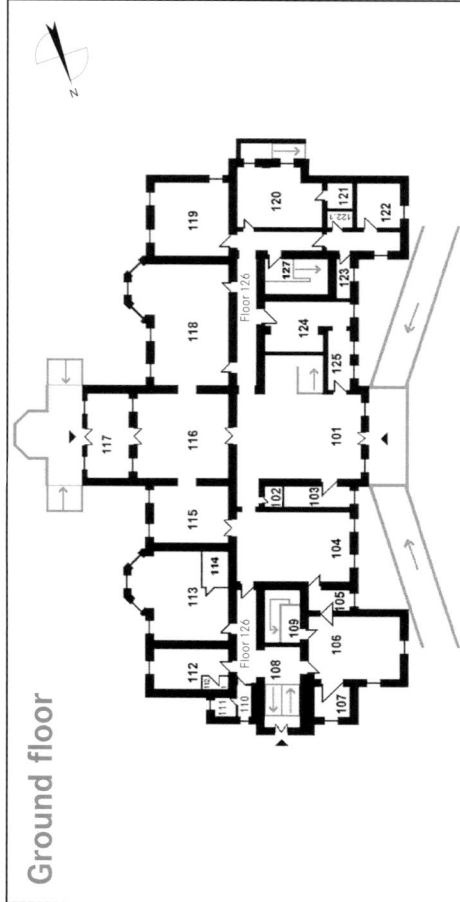

Basement

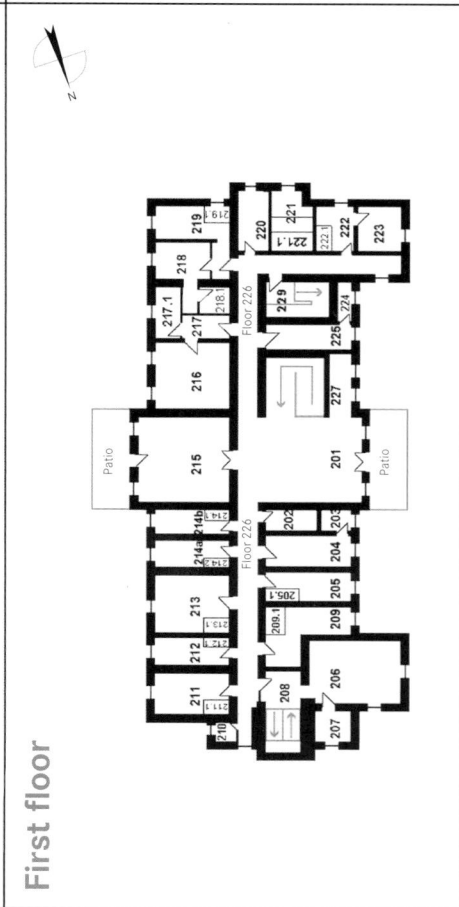

Drink water
intake

☐ Sewage lift station
● Perimeter drain gully
○ Rainwater gully

Site of trickling
filter sewage treatment system

First floor

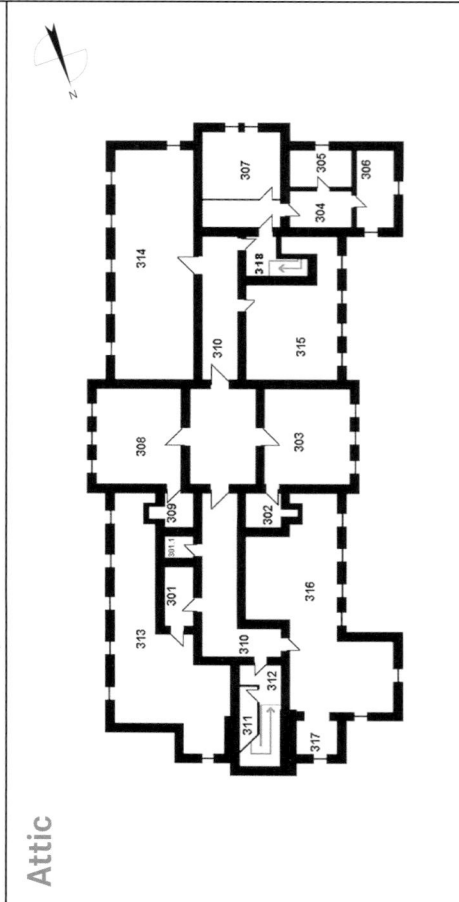

Patio

Patio

Maintenance care

Frequency: See work guidance

First performed: May 05

Servicing

Frequency:

First performed:

Task	Work guidance	Additional comments	Material description/supply source
	See work guidance		See work guidance
	May 05		May 05
Sewage system, lift station, drains	Twice a year (inspect and clean).	Take out maintenance contract with specialist contractor.	Sewage lift station: type Sanimat 1002D/HD. Additional pump: submersible pump U 3 K Niro.
Phosphate additives for sewage treatment, cleaning of trickling filter systems	Regularly check fill level of container, four times a year, as part of servicing, visually inspect (establish control data and quantities).	Take out maintenance contract with specialist contractor.	Trickling filter sewage treatment plant: type Bio-Clear. Enter maintenance contractor with address, tel/fax, e-mail and internet address.
External components, grease interceptors, grease drain in basement	Monthly servicing.	Take out maintenance contract with specialist contractor.	Enter maintenance contractor with address, tel/fax, e-mail and internet address.
Cleaning of septic tank	Frequency as required by statutory provisions: open, rinse, add materials, close.	Take out maintenance contract with specialist contractor.	Enter maintenance contractor with address, tel/fax, e-mail and internet address.
Cleaning of trickling filter sewage treatment system	Clean trickling filter sewage treatment system, manholes, inspect sewage irrigation field. Visually inspect, open, rinse, close, establish control data and quantities. Four times a year.	Take out maintenance contract with specialist contractor.	Enter maintenance contractor with address, tel/fax, e-mail and internet address.
Water pipes	Rinse perimeter drains, rainwater drains, keep inlet structure clear. Visually inspect, open, rinse, close. Once a year. Regularly operate main cocks (approx. four times a year) to prevent them from seizing up.	Take out maintenance contract with specialist contractor. By caretaker. For sanitary inspection and servicing plan (hot/cold water), see annex.	Enter maintenance contractor with address, tel/fax, e-mail and internet address.
Work location	**Basement and external areas** **Sewage lift station in basement, space 015**		
Cross-references	Electrical installations, sheet no. 7/1-7/2 (switch cabinet for sewage treatment system)		
Issued on: 02.03.2007 Name: Frisch	**1st update on: 15.07.2008** Changes: Name: Klemisch	**2nd update on:** Changes: Name:	**3rd update on:** Changes: Name:

Ground floor

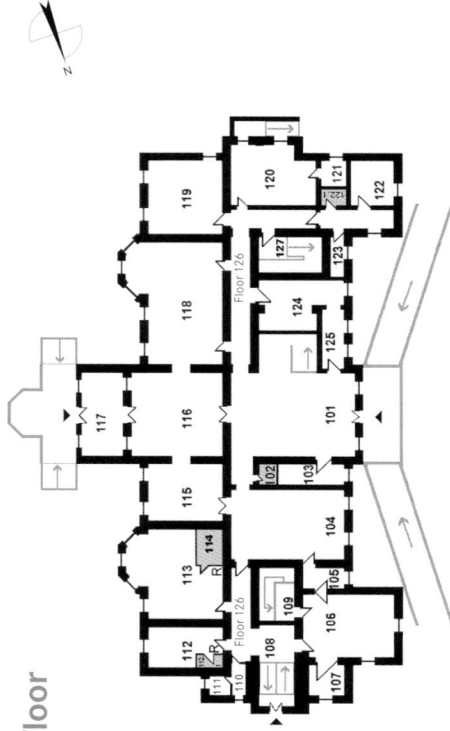

Attic

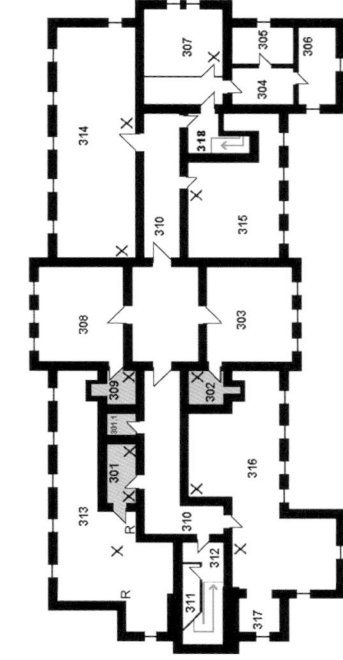

X Chimney vents

Basement

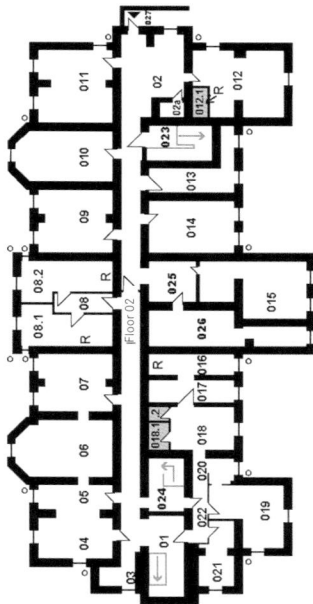

o Downpipes
R Access openings

First floor

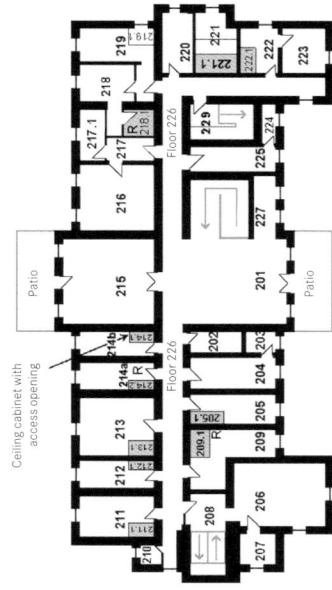

Ceiling cabinet with
access opening

Inspection

Frequency: Twice a year

First performed: May 05

Task	Work guidance	Additional comments	Material description/supply source
Inspection of disused chimneys	Inspect disused chimneys for signs of condensation damage, check chimney vents.	By specialist contractor.	
Inspection of downpipes	Check for soiling and proper flow.	By caretaker.	
Inspection of extractor fans (internal bathrooms)	Check for soiling and smooth operation.	By caretaker.	
Access openings in ventilation ducts (see Section 14/2)	Open as necessary (e. g. in emergencies).	By caretaker or specialist contractor.	
Work location	**Building services (access openings in ventilation ducts), downpipes** **Chimney vents in attic** **Spaces with extractor fans: B 018.1–2, 012.1; GF 102, 112.1, 114, 122.1;** **1F 205.1, 209.1, 211.1, 212.1, 214.1–2, 218.1, 221.1, 222.1; A 301, 301.1, 302, 309**		

Cross-references	Ventilation, sheet no. 14/1–14/2; Heating, sheet no. 8/1–8/2; External drainage, sheet no. 11/3–11/4		
Issued on: 02.03.2007	**1st update on: 11.10.2007**	**2nd update on: 15.07.2008**	**3rd update on:**
	Changes:	Changes:	Changes:
Name: Miethe, Plätzer	Name: Frisch	Name: Klemisch	Name:

Ground floor

119 120 121 122 123 127 124 125
118 117 116 101 102 103 104
115 114 113 112 111 110 109 108 107 106 105
Floor 126

Switch cabinet ventilation

Attic

307 305 306 304 318 314 310 315 308 303 309 302 301 316 313 310 312 311 317

Sauna extract
Wine bar extract

Basement

011 02 012 010 09 013 014 08.2 08 025 015 08.1 026 016 07 017 06 018.2 018 05 020 019 04 022 024 01 03 021 027 023 012.1

Floor 02

Refrigerated store

▲ Fresh air
✗ Switch cabinet ventilation

First floor

219 219.1 220 221 222 218 221.1 222.1 223 217.1 217 218.1 229 224 216 225 215 227 201 214a 214b 202 203 213 214.1 205.1 204 205 212 213.1 209.1 209 211 212.1 208 206 210 211.1 207

Patio
Patio
Floor 226
Floor 226

	Frequency:		
	First performed:		
	See work guidance	Frequency:	
	May 05	First performed:	
		May 05	May 05

Task	Work guidance	Additional comments	Material description/supply source
Inspection of extractor fans in internal bathrooms (s. 13/2)	Check fans for soiling and smooth operation. Four times a year.	By caretaker.	
Cleaning of filters	If red indicator lights up, open lid, remove filter unit, clean in dishwasher and re-insert into device.	By caretaker.	Reusable filter. Enter supply source with address, tel/fax, e-mail and internet address.
Inspection of ventilation system (incl. air treatment systems)	Inspect twice a year. Replace filter unit for supply air in air-conditioner as required.	Take out maintenance contract with specialist contractor.	Bag-type filter. Enter supply source with address, tel/fax, e-mail and internet address.
Cleaning of ventilation system	Clean ventilation ducts and plenums, dampers, valves and motors, once a year.	Take out maintenance contract with specialist contractor. For ventilation system servicing plan, see annex.	Enter maintenance contractor with address, tel/fax, e-mail and internet address.
Work location	Ventilation system components in basement, spaces 013, 016–022, in attic, spaces 315, 316 and kitchen, space 106 Air treatment systems in basement, space 012 (sauna), space 04–06 (wine bar)		
Cross-references			
Issued on: 02.03.2007 Name: Miethe, Plätzer	**1st update on: 11.10.200/** Changes: Name: Frisch	**2nd update on: 15.07.2008** Changes: Name: Klemisch	**3rd update on:** Changes: Name:

Ground floor

Attic

Basement

First floor

Cleaning		Servicing	
Frequency:		Frequency:	Scc work guidance
First performed:	See work guidance	First performed:	May 05
	May 05		

Care	Work guidance	Additional comments	Material description/supply source
Gilding of fibrous plaster ceilings	Clean and touch up as part of ceiling painting.	Only to be performed by specialist contractor. Duty to notify owner.	Enter contractor's address, tel/fax, e-mail and internet address.
Maintenance			
Fibrous plaster ceilings	Paint only with spray equipment (to retain profiling). Colour RAL 9010 (white).	Maintenance painting and repainting only in consultation with owner. Call in specialist contractor.	Colour RAL 9010 (white). Specialist contractor as specified above.
Fire protective coating	Do not overcoat, touch up immediately in case of damage, and check intensively at least every five years.	Only to be performed by specialist contractor. Duty to notify owner.	Enter make, RAL 9010.
Internal woodwork	Maintenance painting and repainting; maintenance painting every 4 years, stripping and repainting every 8 years, touch up any damage immediately.	Only to be performed by specialist contractor. Duty to notify owner.	Enter make. High-build woodstain, colour 048, rosewood. Enter maintenance contractor, supply source.
Heat emitters and flow/return pipes	Maintenance painting and repainting.	Only to be performed by specialist contractor.	Colour: RAL 9010, enter specialist contractor.
Internal walls/ceilings generally	Painting every 3 years, wallpapering every 6 years.	To be performed by specialist contractor where appropriate. Duty to notify owner.	Colour: walls RAL 9002; ceilings RAL 9010. Enter supply source. Enter details of any specialist contractor, with address, tel/fax, e-mail and internet address.

Work location	Internal walls and ceilings, gilding to hall 118	
Cross-references	Heat emitters, sheet no. 8.2	
Issued on: 02.03.2007	**1st update on: 12.09.2007**	**2nd update on: 15.07.2008**	**3rd update on:**
	Changes:	Changes:	Changes:
Name: Miethe, Plätzer	Name: Frisch	Name: Klemisch	Name:

Eastern façade

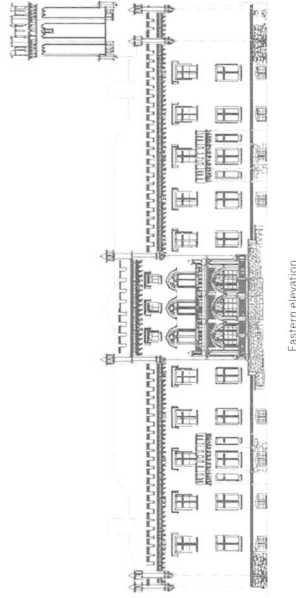

Eastern elevation

Western façade

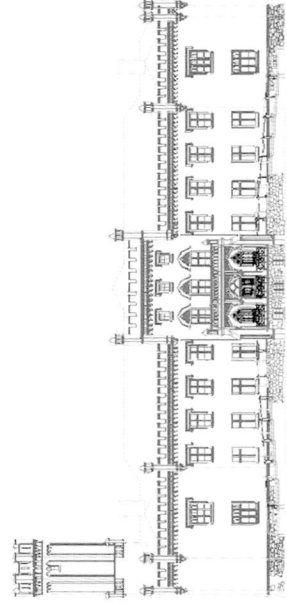

Western elevation

Northern façade

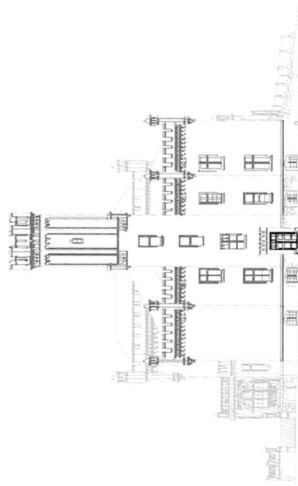

Southern façade

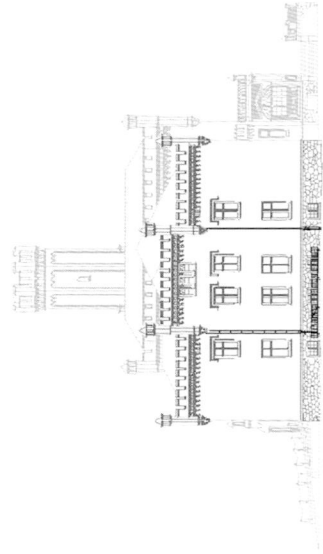

Inspection	Maintenance painting (overpainting)	
Frequency: Monthly	Frequency:	Every 10 years
First performed: May 05	First performed:	May 05

Task	Work guidance	Additional comments	Material description / supply source
Inspection for damp, efflorescence, flaking paint and discoloration	Visually inspect approx. once a year.	By caretaker. Duty to notify owner in case of damage.	Lime / cement render (with Roman cement) av. 20 mm thick, floated). Coating system: primer for silicate emulsion paint, colour 9268, enter make.
Inspection of window sills, door / entrance areas, cornices and string courses for efflorescence and flaking	Visually inspect after heavy rainfall or once a month.	By caretaker. Duty to notify owner in case of damage.	Plinth area: granite masonry. Pointed arch frieze and corbelled cornice: picked out in single colour 9271.
Maintenance painting (overpainting)	Perform works in consultation with paint manufacturer (KEIMFARBEN), paint full surface. Maintenance approx. every 10 years. Touch up immediately in case of damage.	Consultation with owner required. Call in specialist contractor.	Enter supply source, specialist contractor with address, tel / fax, e-mail and internet address.

Work location	**External wall, façade**		
Cross-references	Windows, sheet no. 3 / 1–3 / 4; Plinths sheet no. 16 / 3		
Issued on: 02.03.2007 Name: Miethe, Plätzer	**1st update on: 15.07.2008** Changes: Name: Frisch	**2nd update on:** Changes: Name:	**3rd update on:** Changes: Name:

Property:

Work card

321

External walls, plinths
Sheet no. 16 / 3

Northern façade

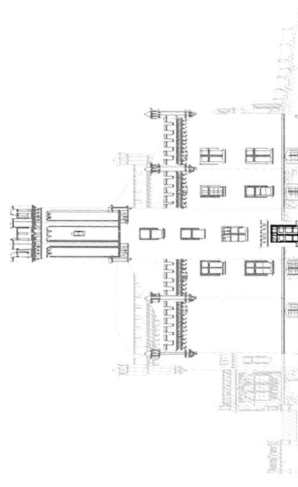

Eastern façade

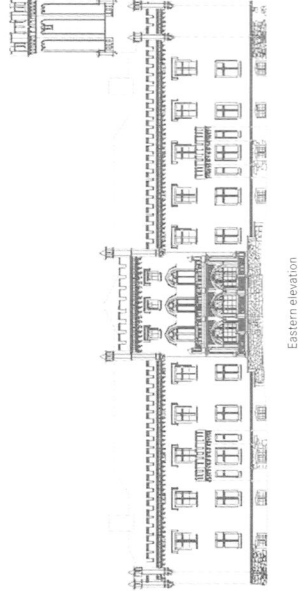

Southern façade

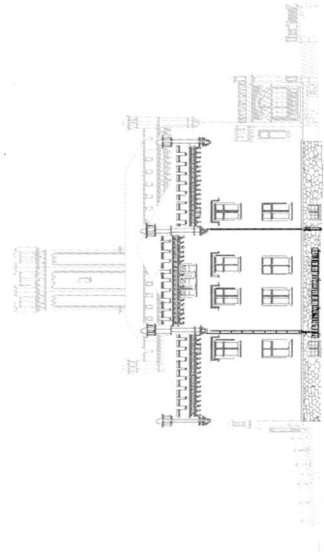

Western façade

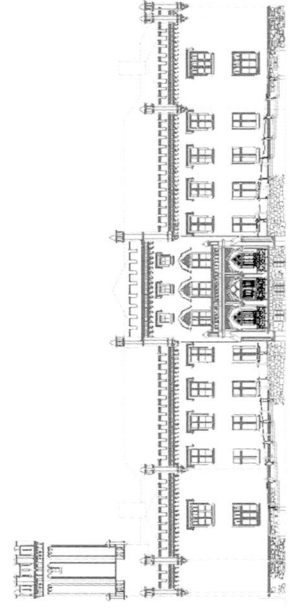

External walls, plinths

Routine cleaning	Servicing	
Frequency: Twice a year	Frequency: See work guidance	
First performed: May 05	First performed: May 05	

Task	Work guidance	Additional comments	Material description / supply source
Cleaning	Sweep and hose down twice a year. Use mechanical brush equipment in case of heavy soiling.	By caretaker.	Plinth area: granite masonry.
Maintenance	Repair in case of damage or unsightly weathering.	Duty to notify owner. To be performed by suitable specialist contractor.	
Inspection	Inspection for discoloration and damp by caretaker, after heavy rainfall and once a month.		

Work location	Plinth area of façade

Cross-references			
Issued on: 02.03.2007	1st update on: 16.08.2007	2nd update on: 15.07.2008	3rd update on:
Name: Miethe, Plätzer	Changes: Name: Frisch	Changes: Name: Klemisch	Changes: Name:

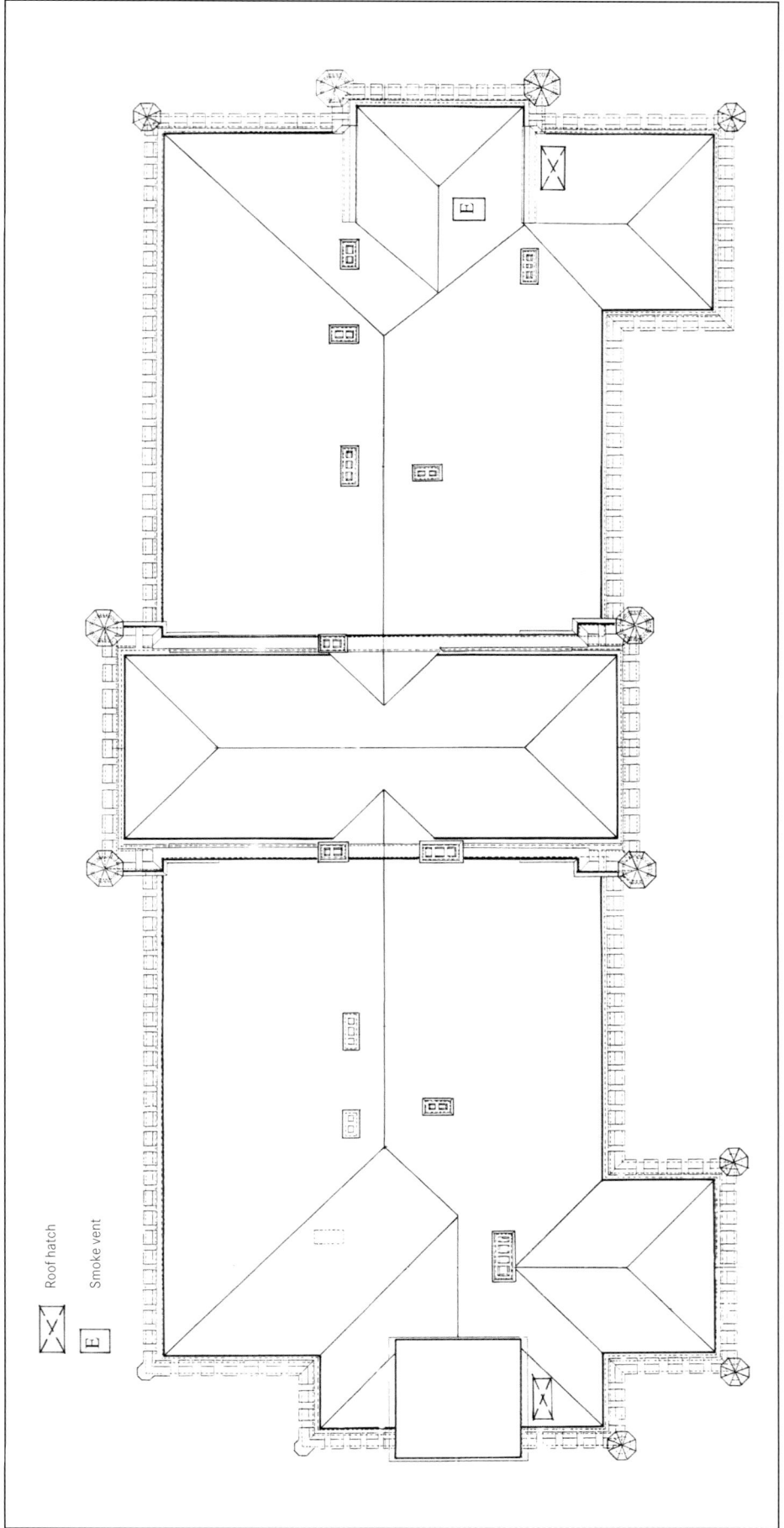

Roof hatch

E Smoke vent

Inspection		Roof cleaning	
Frequency: Once a month		Frequency:	Once a year
First performed: May 05		First performed:	May 05

Task	Work guidance	Additional comments	Material description/supply source
Roof covering	Visually inspect slating through roof hatch, from tower and attic for watertightness and defects, e.g. missing or damaged slates, once a month.	Inspection by caretaker. To be repaired by specialist contractor. Duty to notify owner.	English slate: Greaves Portmadoc Slate, size 61.0 x 30.5 cm, 4–5 mm thick. For repairs: store remaining slates in loft.
Crenellations	Check for cracks, water infiltration through joints and organic growth.	To be repaired by specialist contractor. Duty to notify owner in case of abnormalities.	0.7–1.0 mm zinc sheet flashing to crenellations; jointing with zinc solder. Enter supply source. Enter details of any specialist contractor, with address, tel./fax, e-mail and internet address.
Lightning protection	Visually inspect lightning protection system (fixings). Refasten/secure hollow slates where necessary. Inspect for any signs of damage. System serviced once a year.	By caretaker or specialist contractor. Duty to notify owner in case of abnormalities. Take out maintenance contract with specialist contractor.	Enter maintenance contractor.
Inspection for watertightness of chimneys	Check for new traces of water on floor, every 14 days or after heavy rainfall. Inspect disused chimneys for signs of condensation damage etc.; twice a year.	By caretaker. In case of water damage, immediately notify owner and arrange for roof/chimney junctions to be checked by specialist contractor. Take out maintenance contract with specialist contractor or arrange for inspection by chimneysweep.	Enter details of any specialist contractor, with address, tel/fax, e-mail and internet address.

Work location	Roof		
Cross-references	Drainage, sheet no. 11/1-11/4		
Issued on: **02.03.2007**	1st update on: **12.09.2007**	2nd update on: **15.07.2008**	3rd update on:
Name: Miethe, Plätzer	Changes: Name: Frisch	Changes: Name: Klemisch	Changes: Name:

Sheet no. 17/2

Property: **Work card**

328 **External works**

Sheet no. 18/1

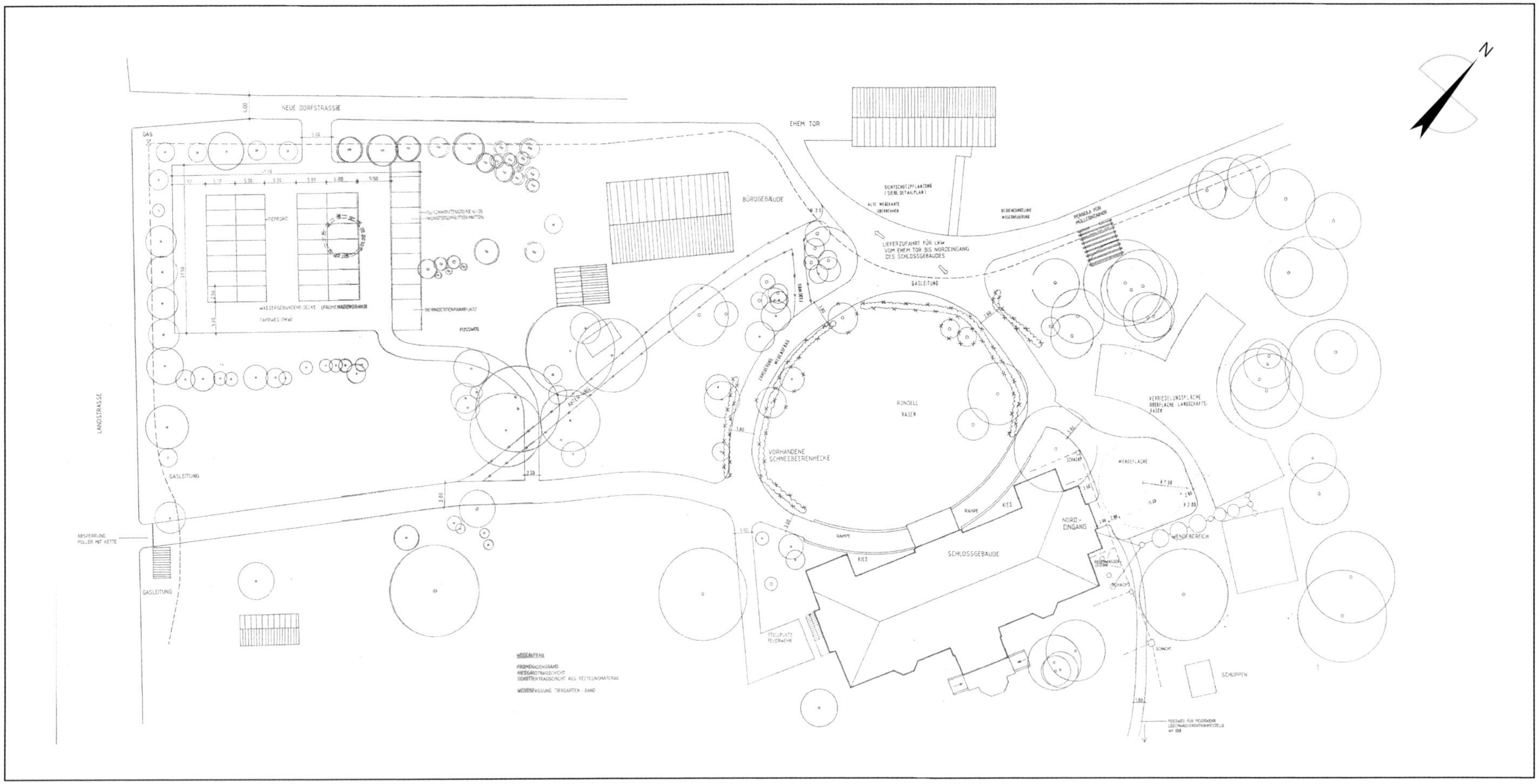

Maintenance care

Frequency:

First performed:

Servicing

Frequency:

First performed:

	As required	As required
	May 05	May 05

Task	Work guidance	Additional comments	Material description/supply source
Rinsing of perimeter drains, rainwater drains, keeping inlet structures clear	Visually inspect. Open, rinse, close once a year.	To be performed by specialist contractor. Take out maintenance contract.	Enter maintenance contractor with address, tel/fax, e-mail and internet address
Driveway	Monitor spread of gravel (rutting) and proper drainage, ensure snow clearance and permanent access for rescue vehicles. Inspect and clean, level out gravel covering, after heavy rainfall, otherwise once a month.	Levelling out by caretaker. By gardener or caretaker.	
Pathways, car park	Inspect paths to building, incl. car park, for trip hazards or other damage (potholes, frost damage). Keep pathways clear throughout year (leaf fall). Perform obligatory snow clearance and gritting.	Call in specialist contractor where necessary following consultation with owner. By caretaker.	
Firefighting water	Fire service access route from Dorfstraße to lake and firefighting water draw-off point to be kept clear, in consultation with municipal authorities.	In accordance with statutory provisions and during winter every day or as required.	
External stairways	Check for damaged, worn steps.	Call in specialist contractor where necessary following consultation with owner.	
Planting	Inspect trees and bushes for stability and potential hazards from falling rotten branches, twice a year. Annual pruning.	Inspection by caretaker. Pruning by specialist contractor. Consultation with owner required.	

Work location	External areas		
Cross-references			
Issued on: 02.03.2007	1st update on: 12.09.2007	2nd update on: 15.07.2008	3rd update on:
	Changes:	Changes:	Changes:
Name: Miethe, Plätzer	Name: Frisch	Name: Klemisch	Name:

Sheet no. 18 / 2

8 Example of Condition Survey spreadsheets

Section 6 describes how to prepare the documentation for, and conduct routine inspections of, the building condition. The sample User Maintenance Instructions presented in Section 7 are complemented here by a practical example of the documents used for Condition Surveys.

The preparation of Condition Survey documents, particularly for historic buildings, requires specialist knowledge. In principle, the procedure described is applicable to any facility type – including residential and administrative properties, factories, sports centres or shopping malls. The drafting of Building Maintenance Logbooks and performance of the associated Condition Surveys, including preparation and follow-up, offer architects and engineers as yet largely untapped opportunities to offer new services.

By illustrating the procedure outlined in Section 6, the following examples are intended to help readers prepare their own, property-specific Condition Survey documents.

The specimen documents are based on a real building, Schloss Reichenow, which is located in the German Federal State of Brandenburg and described in greater detail in Section 9.

Condition Survey, contents

Condition Survey from outside

Condition Survey from inside

Condition Survey from inside

Work cards for individual element groups

Element group: Northern façade

Code 6.3.1.1

Survey date:		
Officer:		
Cross-references:		
Scale: none		

Property:

Date:

Signature:

Damage type / cause	Findings	n.f.	Damage type / cause	Findings	n.f.
Damage due to damp			Damage at joints / junctions		
Efflorescence			– Windows		
Chalking			– Window sills		
Flaking, blistering			– Doors		
Peeling from base			– Cornices / string courses		
Discolouration			– Roofing		
Infection			– Downpipes, gargoyles, gutters		
Organic growth					
Damage due to soiling			Clogging at base of downpipes? Access opening		
Damage due to external mechanical action			Jointing intact?		
Structural damage / cracking			Stripping and repainting necessary?		
			others		

Property: Date: Signature:

Element group: Northern façade

Text page / damage record

Code 6.3.1.1

Element:

Findings no.	Description	Necessary action	Initiated date / signature	Completed date / signature

Property: Date: Signature:

128

Element:

Findings no.	Description	Necessary action	Initiated date / signature	Completed date / signature

Property: Date: Signature:

Element group: Eastern façade

Survey date:			
Officer:		Cross-references:	
			Scale: none

Property: Date: Signature:

Element group: Eastern façade

Damage type/cause	Findings	n.f.	Damage type/cause	Findings	n.f.
Damage due to damp			Damage at joints/junctions		others
Efflorescence			– Windows		
Chalking			– Window sills		
Flaking, blistering			– Doors		
Peeling from base			– Cornices / string courses		
Discolouration			– Roofing		
Infection			– Downpipes, gargoyles, gutters		
Organic growth					
Damage due to soiling			Clogging at base of downpipes? Access opening		
Damage due to external mechanical action			Jointing intact?		
Structural damage / cracking			Stripping and repainting necessary?		

Property: Date: Signature:

Text page / damage record

Code 6.3.1.2

Element:

Findings no.	Description	Necessary action	Initiated date / signature	Completed date / signature

Property:

Date:

Signature:

Element group: Eastern façade

Text page / damage record

Code 6.3.1.2

Element:

Findings no.	Description	Necessary action	Initiated date / signature	Completed date / signature

Property:

Date:

Signature:

Element group: Southern façade

Code 6.3.1.3

Survey date:

Officer:

Cross-references:

Scale: none

Property:

Date:

Signature:

Damage type/cause	Findings	n.f.	Damage type/cause	Findings	n.f.
Damage due to damp			others		
Efflorescence					
Chalking					
Flaking, blistering					
Peeling from base					
Discolouration					
Infection					
Organic growth					
Damage due to soiling					
Damage due to external mechanical action					
Structural damage/cracking					

Damage type/cause	Findings	n.f.
Damage at joints/junctions		
– Windows		
– Window sills		
– Doors		
– Cornices/string courses		
– Roofing		
– Downpipes, gargoyles, gutters		
Clogging at base of downpipes? Access opening		
Jointing intact?		
Stripping and repainting necessary?		

Property:

Date:

Signature:

Element group: Southern façade

Text page / damage record

Code 6.3.1.3

Element:

Findings no.	Description	Necessary action	Initiated date / signature	Completed date / signature

Property: Date: Signature:

Element group: Southern façade   Text page / damage record   Code 6.3.1.3

Element:

Findings no.	Description	Necessary action	Initiated date / signature	Completed date / signature

Property: Date: Signature:

Element group: Western façade

Code 6.3.1.4

Survey date:	
Officer:	
Cross-references:	
Scale: none	

Property:

Signature:

Date:

Element group: Western façade

Code 6.3.1.4

Damage type / cause	Findings	n.f.	Damage type / cause	Findings	n.f.
Damage due to damp			Damage at joints / junctions		
Efflorescence			– Windows		
Chalking			– Window sills		
Flaking, blistering			– Doors		
Peeling from base			– Cornices / string courses		
Discolouration			– Roofing		
Infection			– Downpipes, gargoyles, gutters		
Organic growth					
Damage due to soiling			Clogging at base of downpipes? Access opening		
Damage due to external mechanical action			Jointing intact?		
Structural damage / cracking			Stripping and repainting necessary?		
			others		

Property:

Date:

Signature:

139

Element group: Western façade

Text page / damage record

Code 6.3.1.4

Element:

Findings no.	Description	Necessary action	Initiated date / signature	Completed date / signature

Property: Date: Signature:

Element group: Western façade

Text page / damage record

Code 6.3.1.4

Element:

Findings no.	Description	Necessary action	Initiated date / signature	Completed date / signature

Property:

Date:

Signature:

Element group: Roof plan

Code 6.3.2

N

Roof hatch

E Smoke vent

E

E

Survey date:	
Officer:	
Cross-references:	See also 6.3.3 Attic/loft and structural system
Scale: none	

Property: Signature: Date:

Element group: Roof plan

Roof section	Roof covering		Chimney stacks		Valleys, ridges, hips		Dormer windows, hatches		Walkways, ladders		Roof gutters		Lightning protection		Others
	Waterproof	Damaged	n.f.	Damaged	n.f.	Damaged	n.f.	Damaged	n.f.	Damaged	n.f.	Damaged	n.f.	Damaged	

Property:

Date:

Signature:

Element group: Roof plan

Text page/damage record

Code 6.3.2

Element:

Findings no.	Description	Necessary action	Initiated date/ signature	Completed date/ signature

Property: Date: Signature:

Text page / damage record

Element:

Findings no.	Description	Necessary action	Initiated date / signature	Completed date / signature

Property:

Date:

Signature:

Element group: Attic/loft and structural system

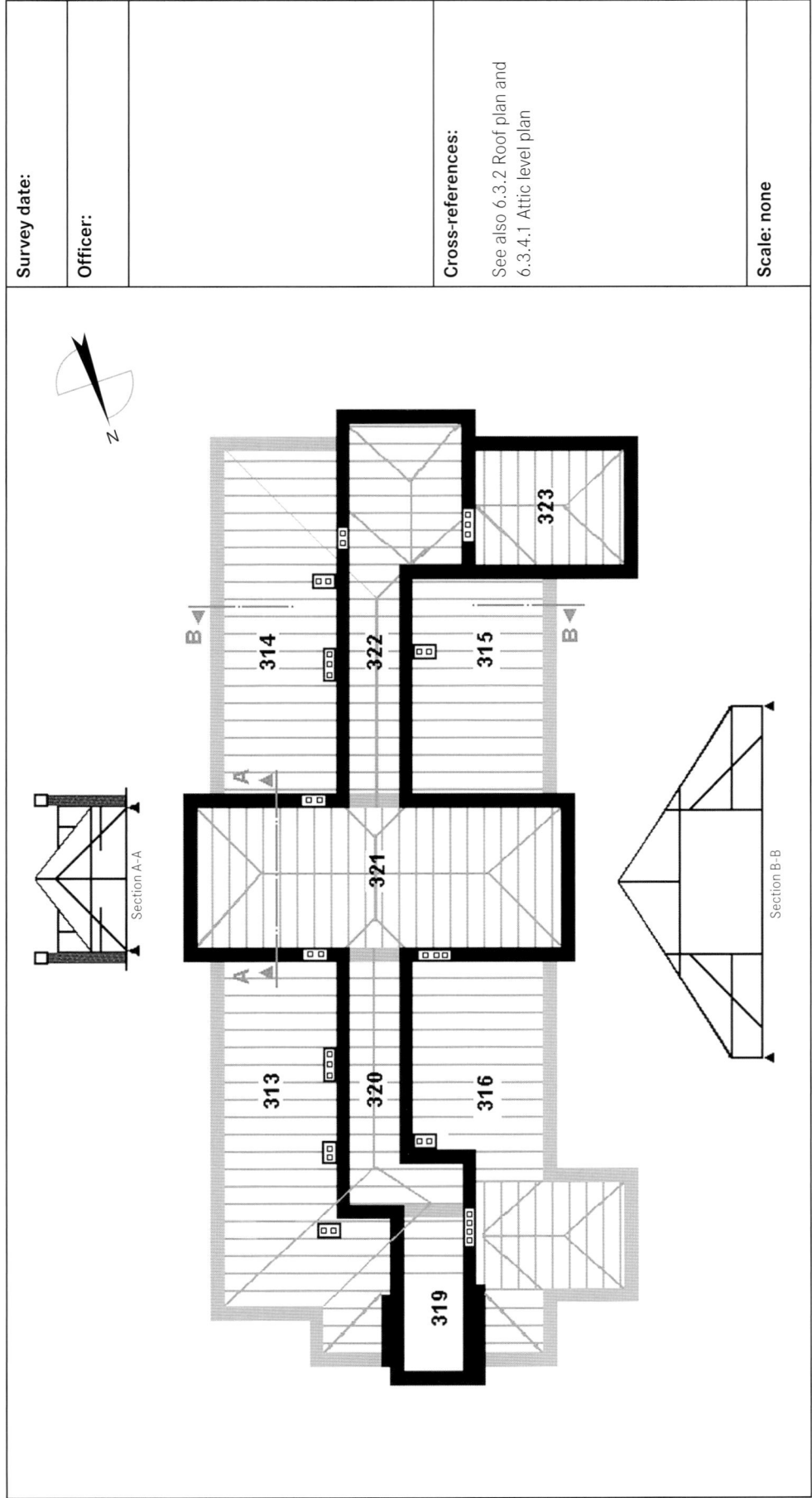

Survey date:	
Officer:	
Cross-references: See also 6.3.2 Roof plan and 6.3.4.1 Attic level plan	
Scale: none	

313

320

316

319

314

322

315

323

321

B

B

A

A

Section A-A

Section B-B

N

Property:

Date:

Signature:

Damage type / cause	Findings	n.f.	Damage type / cause	Findings	n.f.
Structural damage:			Damage due to damp:		
– Rafters / beam ends			– Staining		
– Ventilation of timbers at junctions with masonry / wall plates			– Snow ingress		
– Pugging, pugging boards			– Pugging, pugging boards		
– Collar / tie beams					
– Ridge area			Fungal attack		
– Other structural nodes			Timber infestation, bore dust		
– Floorboards / coverings					
Structural continuity at joints:			Alterations / fixtures		
– Turnbuckles tightened			Insulation / thermal bridging / problem areas		
– Bolts / studs tightened			Loft ventilation		
Improper material storage / overloading					

Property: Date: Signature:

Element:

Findings no.	Description	Necessary action	Initiated date/ signature	Completed date/ signature

Property:

Date:

Signature:

Element:

Findings no.	Description	Necessary action	Initiated date / signature	Completed date / signature

Property:

Date:

Signature:

Element group: Attic level plan

Code 6.3.4.1

| Survey date: |
| Officer: |

Cross-references:

Check stored goods, loads?

Ventilation / environmental conditions?

Nesting places?

See also 6.3.2 Roof plan and 6.3.3 Attic / loft and structural system

Scale: none

Wet areas

Property:

Date:

Signature:

Element group: Attic level plan

Space	Walls		Ceilings		Flooring		Doors		Windows		Fixtures		Services	Others
	Constr.	Finish	Constr.	Finish	Constr.	Finish	Constr.	Finish	Constr.	Finish	Constr.	Finish		

Property: Date: Signature:

Element group: Attic level plan

Text page / damage record

Code 6.3.4.1

Element:

Findings no.	Description	Necessary action	Initiated date / signature	Completed date / signature

Property:

Date:

Signature:

Element:

Findings no.	Description	Necessary action	Initiated date / signature	Completed date / signature

Property: Date: Signature:

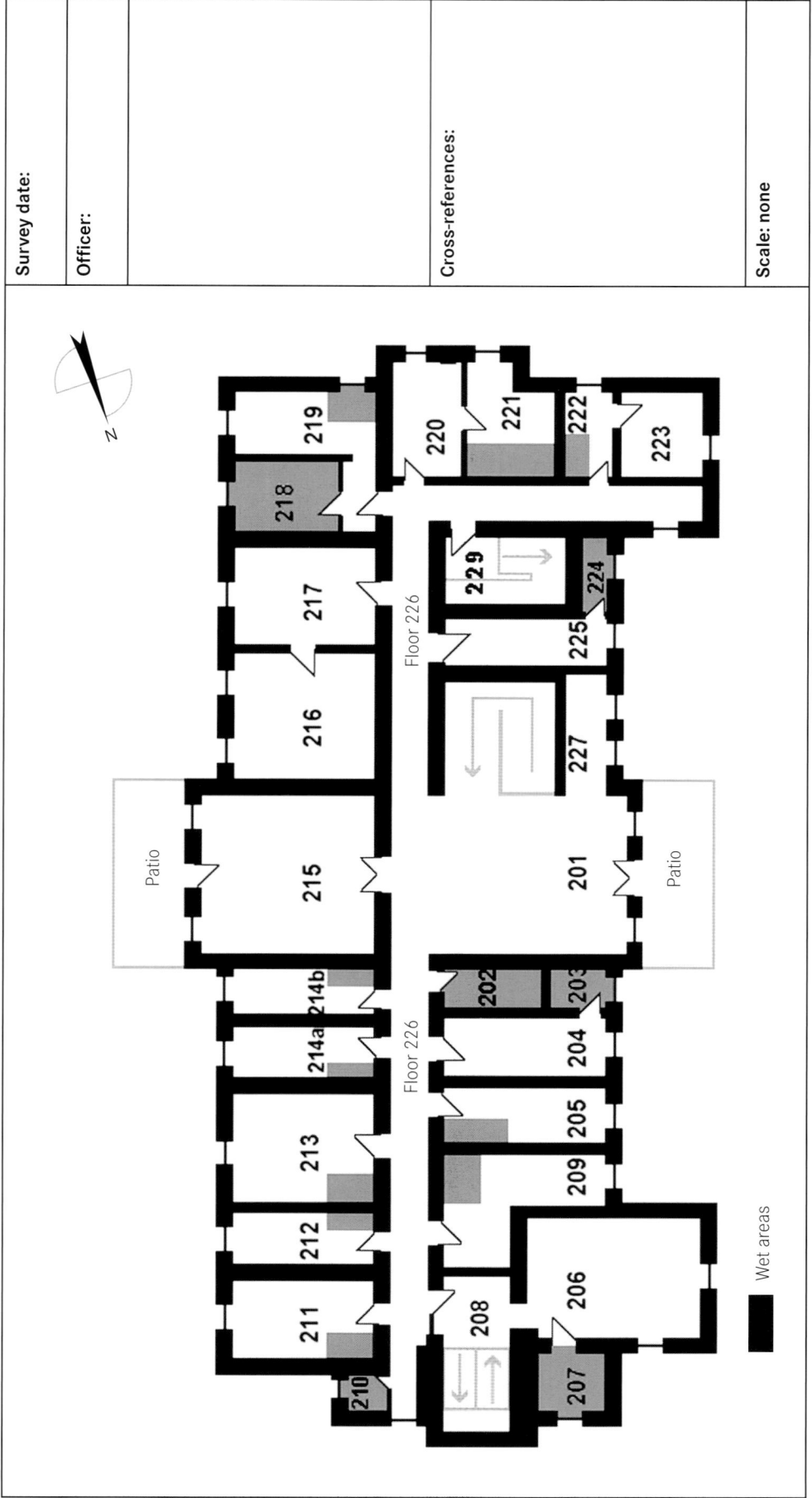

Element group: First floor plan

Code 6.3.4.2

N

219

218

217

216

215

Patio

214a 214b

213

212

211

210

220

221

222

223

229

224

225

227

201

Patio

202

203

204

205

209

206

207

208

Floor 226

Floor 226

Wet areas

Property:

Date:

Signature:

Element group: First floor plan

Space	Walls		Ceilings		Flooring		Doors		Windows		Fixtures		Services	Others
	Constr.	Finish	Constr.	Finish	Constr.	Finish	Constr.	Finish	Constr.	Finish	Constr.	Finish		

Property: Date: Signature:

Element group: First floor plan

Text page / damage record

Code 6.3.4.2

Element:

Findings no.	Description	Necessary action	Initiated date / signature	Completed date / signature

Property:

Date:

Signature:

Element:

Findings no.	Description	Necessary action	Initiated date / signature	Completed date / signature

Property: Date: Signature:

Code 6.3.4.3

Element group: Ground floor plan

| Survey date: |
| Officer: |
| |
| Cross-references: |
| Scale: none |

N

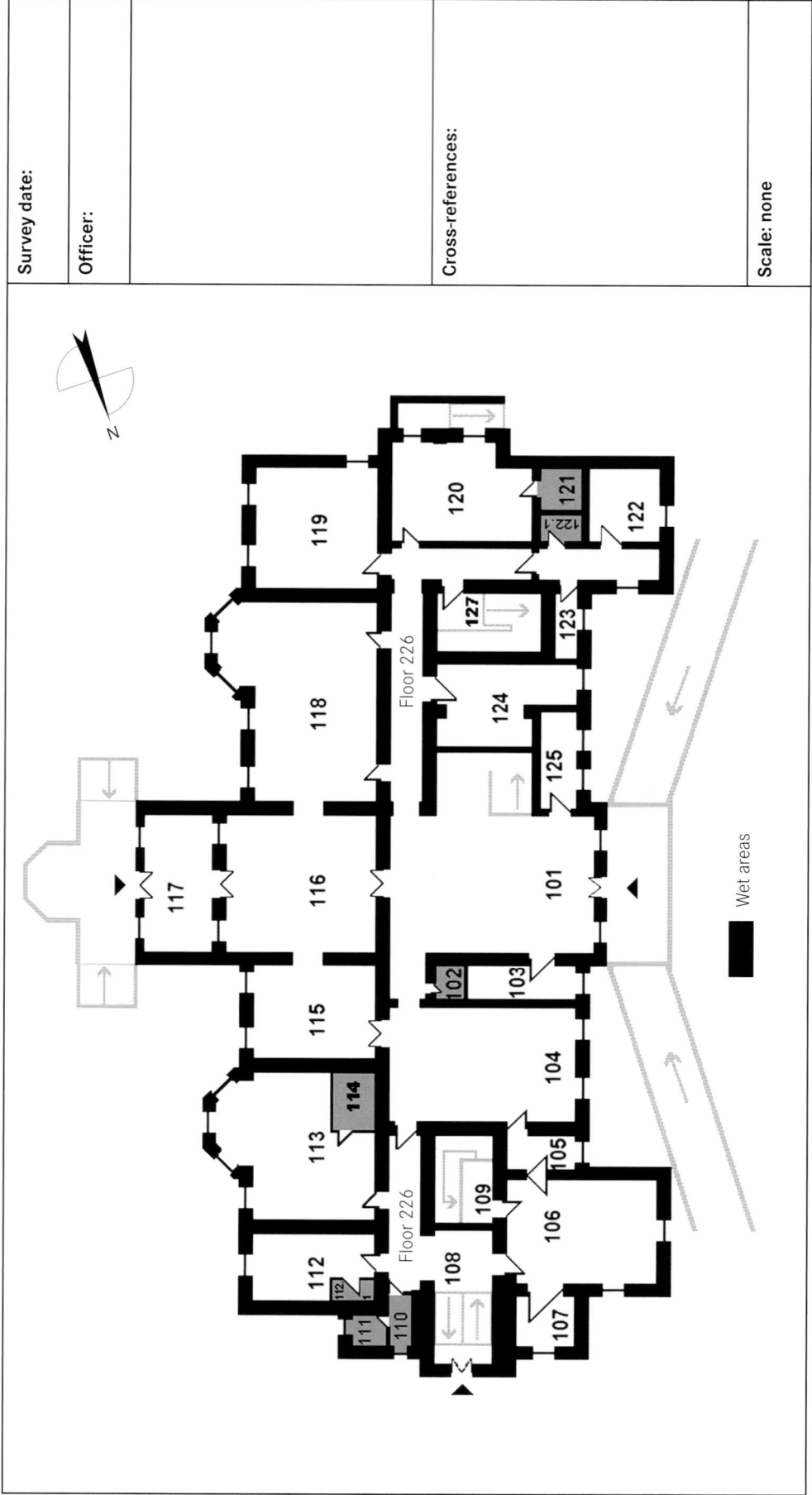

119

120

121

122.1

122

127

123

118

Floor 226

124

125

117

116

101

Wet areas

115

102

103

114

104

113

105

109

112

108

106

111 110

107

Floor 226

Property:

Date:

Signature:

Element group: Ground floor plan

Code 6.3.4.3

Space	Walls		Ceilings		Flooring		Doors		Windows		Fixtures		Services	Others
	Constr.	Finish	Constr.	Finish	Constr.	Finish	Constr.	Finish	Constr.	Finish	Constr.	Finish		

Property:

Date:

Signature:

Element group: Ground floor plan

Text page / damage record

Code 6.3.4.3

Element:

Findings no.	Description	Necessary action	Initiated date / signature	Completed date / signature

Property:

Date:

Signature:

Element:

Findings no.	Description	Necessary action	Initiated date / signature	Completed date / signature

Property: Date: Signature:

Element group: Basement plan

Code 6.3.4.4

| Survey date: |
| Officer: |

Floor 02

■ Wet areas

| Cross-references: |
| Check stored goods, loads? |
| Ventilation / environmental conditions? |
| See also 6.3.5 Basement / waterproofing |
| See also 6.3.1.1 to 6.3.1.4 Façades |

| Scale: none |

Property: Date: Signature:

Element group: Basement plan

Space	Walls		Ceilings		Flooring		Doors		Windows		Fixtures		Services	Others
	Constr.	Finish	Constr.	Finish	Constr.	Finish	Constr.	Finish	Constr.	Finish	Constr.	Finish		

Property: Date: Signature:

Element group: Basement plan

Text page / damage record

Code 6.3.4.4

Element:

Findings no.	Description	Necessary action	Initiated date / signature	Completed date / signature

Property:

Date:

Signature:

Element:

Findings no.	Description	Necessary action	Initiated date / signature	Completed date / signature

Property: Date: Signature:

Element group: Tower levels

Code 6.3.4.5

Section F - F

Section A - A

Section B - B
Stairs
h/b = 18/14
F

Section C - C
Spiral stairs
F h/b = 18/28

Section D - D

Section E - E
Stairs
h/b = 17/22
F

+9.56
+8.50
+8.68

Survey date:

Officer:

Cross-references:

Ventilation / environmental
conditions?

Nesting places?

See also 6.3.2 Roof plan and 6.3.3
Attic / loft and structural system

Scale: none

Signature:

Date:

Property:

Element group: Tower levels

Space	Walls		Ceilings		Flooring		Doors		Windows		Fixtures		Services	Others
	Constr.	Finish	Constr.	Finish	Constr.	Finish	Constr.	Finish	Constr.	Finish	Constr.	Finish		

Property:

Date:

Signature:

Element:

Findings no.	Description	Necessary action	Initiated date/ signature	Completed date/ signature

Property: Date: Signature:

Element:

Findings no.	Description	Necessary action	Initiated date / signature	Completed date / signature

Property: Date: Signature:

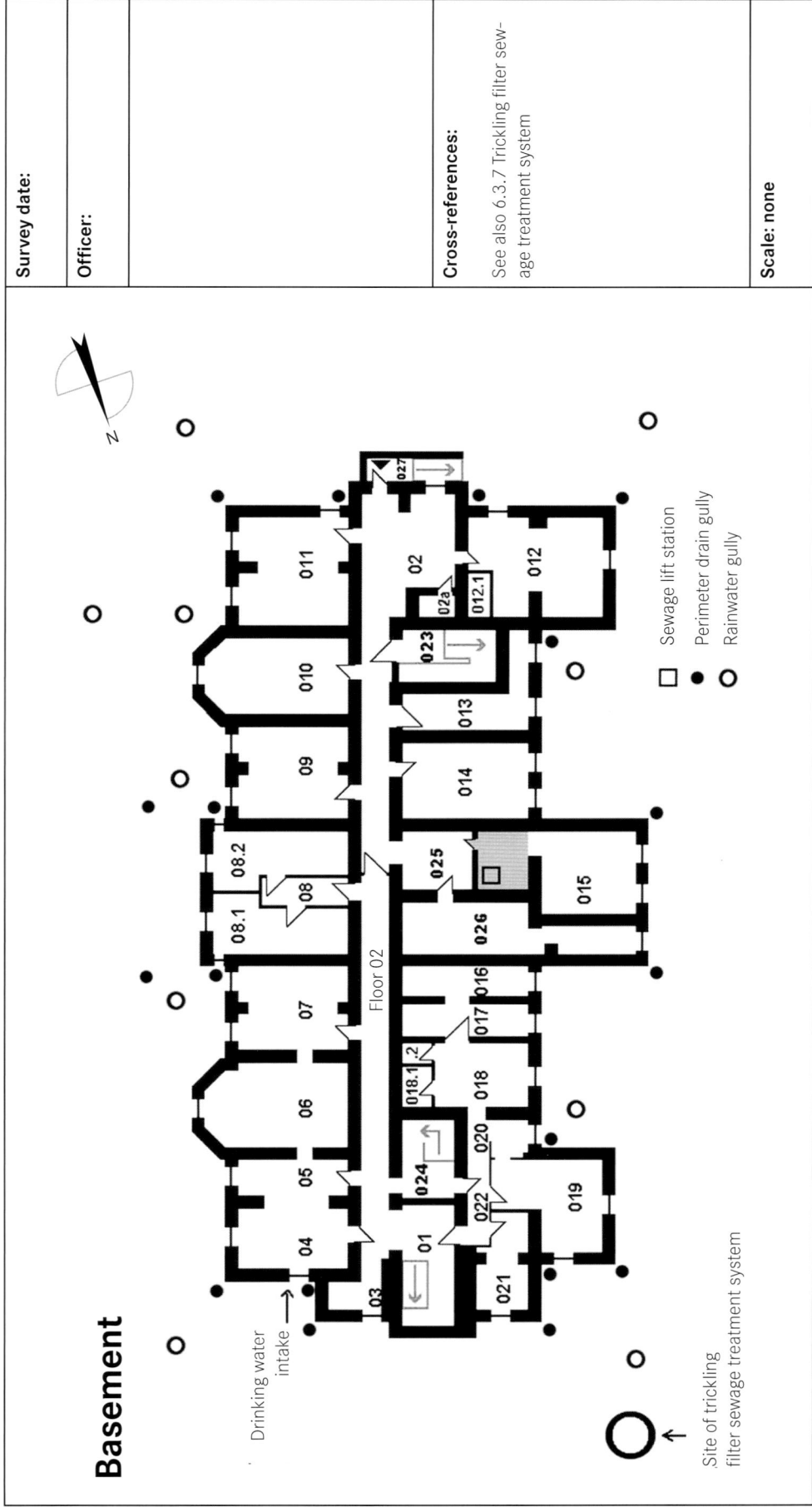

Element group: Basement, structual waterproofing and foundations

Code 6.3.5

Basement

Drinking water intake →

Floor 02

011

010

09

08.2

08.1

08

07

06

05

04

03

01

02

02a

023

012.1

012

013

014

025

026

015

016

017

018

018.1 .2

020

022

019

021

024

027

□ Sewage lift station
● Perimeter drain gully
○ Rainwater gully

Site of trickling
filter sewage treatment system

| Survey date: |
| Officer: |
| Cross-references: |
| See also 6.3.7 Trickling filter sew-age treatment system |
| Scale: none |

Property:

Date:

Signature:

Element group: Basement / waterproofing and foundations

Damage type / cause	Findings	n.f.	Damage type / cause	Findings	n.f.	Damage type / cause	Findings	n.f.
Damage due to damp:			Mechanical damage:			Others		
– Floors / beds			– Penetrations: building intake, pipes and cables					
– Below-ground walls			Structural joints, hydrophilic waterstops					
– Internal walls								
– Doors and windows in below-ground areas			Tide lines on walls (damp)					
– Penetrations: building intake, pipes and cables			Fungal attack					
Drains clear / rinsed			Cracking:					
			– Linear cracking					
Structural joints: cracks, gaps			– Concentrated cracking (centric)					
Signs of settlement			– Area cracking					

Property: Date: Signature:

Element group: Basement/waterproofing and foundations   Text page / damage record   Code 6.3.5

Element:

Findings no.	Description	Necessary action	Initiated date / signature	Completed date / signature

Property: Date: Signature:

Element:

Findings no.	Description	Necessary action	Initiated date / signature	Completed date / signature

Property: Date: Signature:

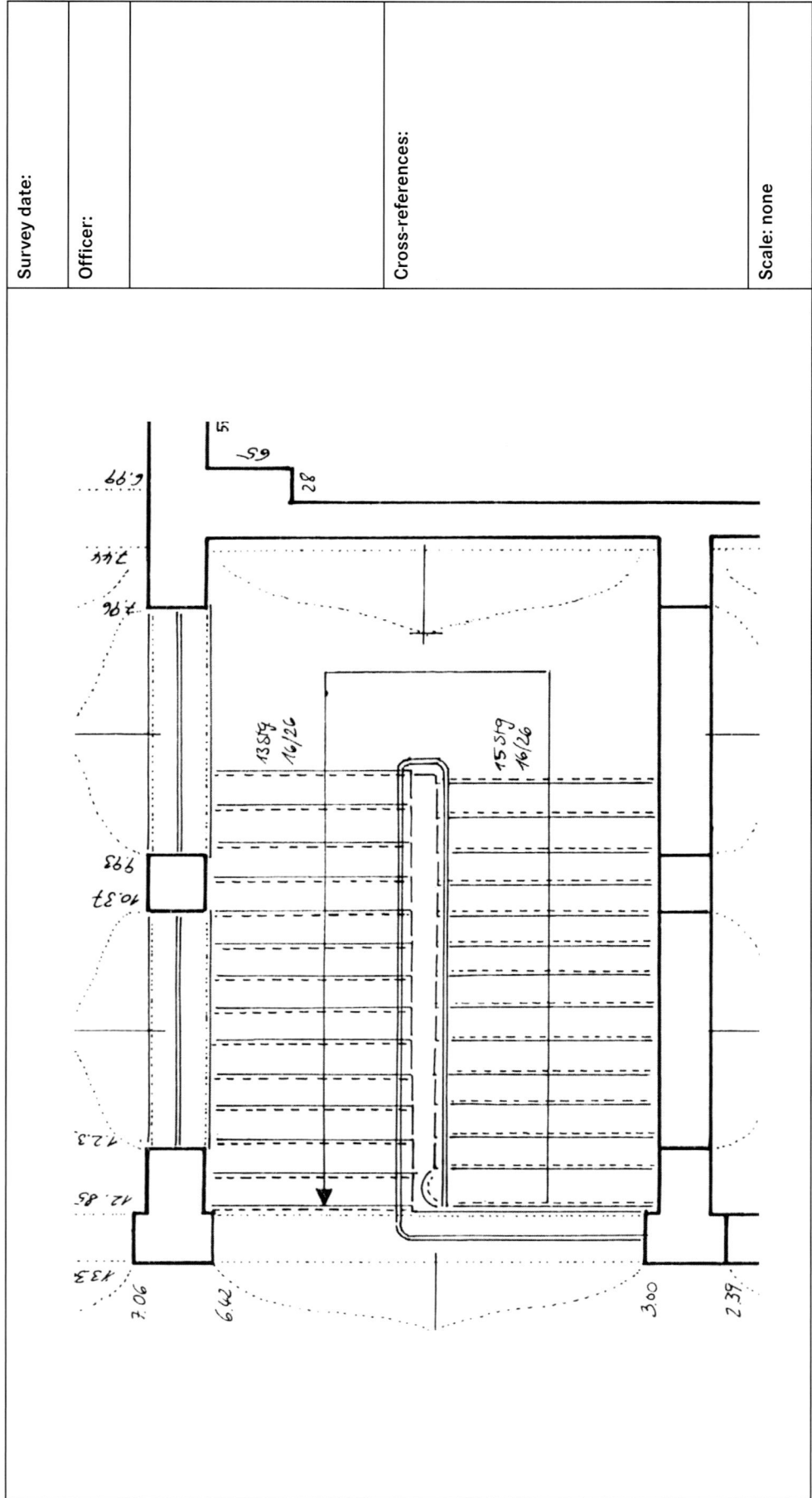

Element group: Main stair

Code 6.3.6.1

Survey date:	
Officer:	
Cross-references:	
Scale: none	

Property:

Signature:

Date:

Element group: Main stair

Stair/	Treads		Risers		Landings		Strings		Covering		Balustrade		Handrail	Others
	Constr.	Finish	Constr.	Finish	Constr.	Finish	Constr.	Finish	Constr.	Finish	Constr.	Finish		

Property:

Date:

Signature:

Element:

Findings no.	Description			Necessary action	Initiated date / signature	Completed date / signature

Property: Date: Signature:

Element:

Findings no.	Description	Necessary action	Initiated date / signature	Completed date / signature

Property: Date: Signature:

Element group: Southern stair

Code 6.3.6.2

Survey date:

Officer:

Cross-references:

Scale: none

Ground-floor plan

Attic plan

Basement plan

First-floor plan

A

1 F

GF

B

Selection A - A

Property:

Date:

Signature:

Element group: Southern stair

Stair/	Treads		Risers		Landings		Strings		Covering		Balustrade		Handrail	Others
	Constr.	Finish	Constr.	Finish	Constr.	Finish	Constr.	Finish	Constr.	Finish	Constr.	Finish		

Property: Date: Signature:

Element:

Findings no.	Description	Necessary action	Initiated date/ signature	Completed date/ signature

Property: Date: Signature:

Element:

Findings no.	Description	Necessary action	Initiated date / signature	Completed date / signature

Property:　　　　　Date:　　　　　Signature:

Element group: Northern stair

Code 6.3.6.3

| Survey date: |
| Officer: |

| Cross-references: |
| Continues into tower stair, see 6.3.6.4 |

| Scale: none |

Signature:

Date:

Property:

Element group: Northern stair

Stair/	Treads		Risers		Landings		Strings		Covering		Balustrade		Handrail	Others
	Constr.	Finish	Constr.	Finish	Constr.	Finish	Constr.	Finish	Constr.	Finish	Constr.	Finish		

Property:

Date:

Signature:

Element group: Northern stair

Text page / damage record

Code 6.3.6.3

Element:

Findings no.	Description	Necessary action	Initiated date / signature	Completed date / signature

Property:

Date:

Signature:

Element:

Findings no.	Description	Necessary action	Initiated date/ signature	Completed date/ signature

Property:

Date:

Signature:

Element group: Tower stair

Code 6.3.6.4

Survey date:		
Officer:		
		Cross-references: Tower levels, see 6.3.4.5 Continues into northern stair, see 6.3.6.3
		Scale: none

Section A – A

Section B – B

Section C – C

Section D – D

Section E – E

Property:

Signature:

Date:

Element group: Tower stair

Stairs	Treads		Risers		Landings		Strings		Covering		Balustrade		Handrail	Others
	Constr.	Finish	Constr.	Finish	Constr.	Finish	Constr.	Finish	Constr.	Finish	Constr.	Finish		

Property: Date: Signature:

Text page / damage record

Element:

Findings no.	Description	Necessary action	Initiated date / signature	Completed date / signature

Property:

Date:

Signature:

Element:

Findings no.	Description	Necessary action	Initiated date / signature	Completed date / signature

Property: Date: Signature:

Element group: Special elements – trickling filter sewage treatment system
and sewage irrigation field

Code 6.3.7

Survey date:
Officer:
Cross-references: See 6.3.5 Check servicing reports and disposal certificates
Scale: none

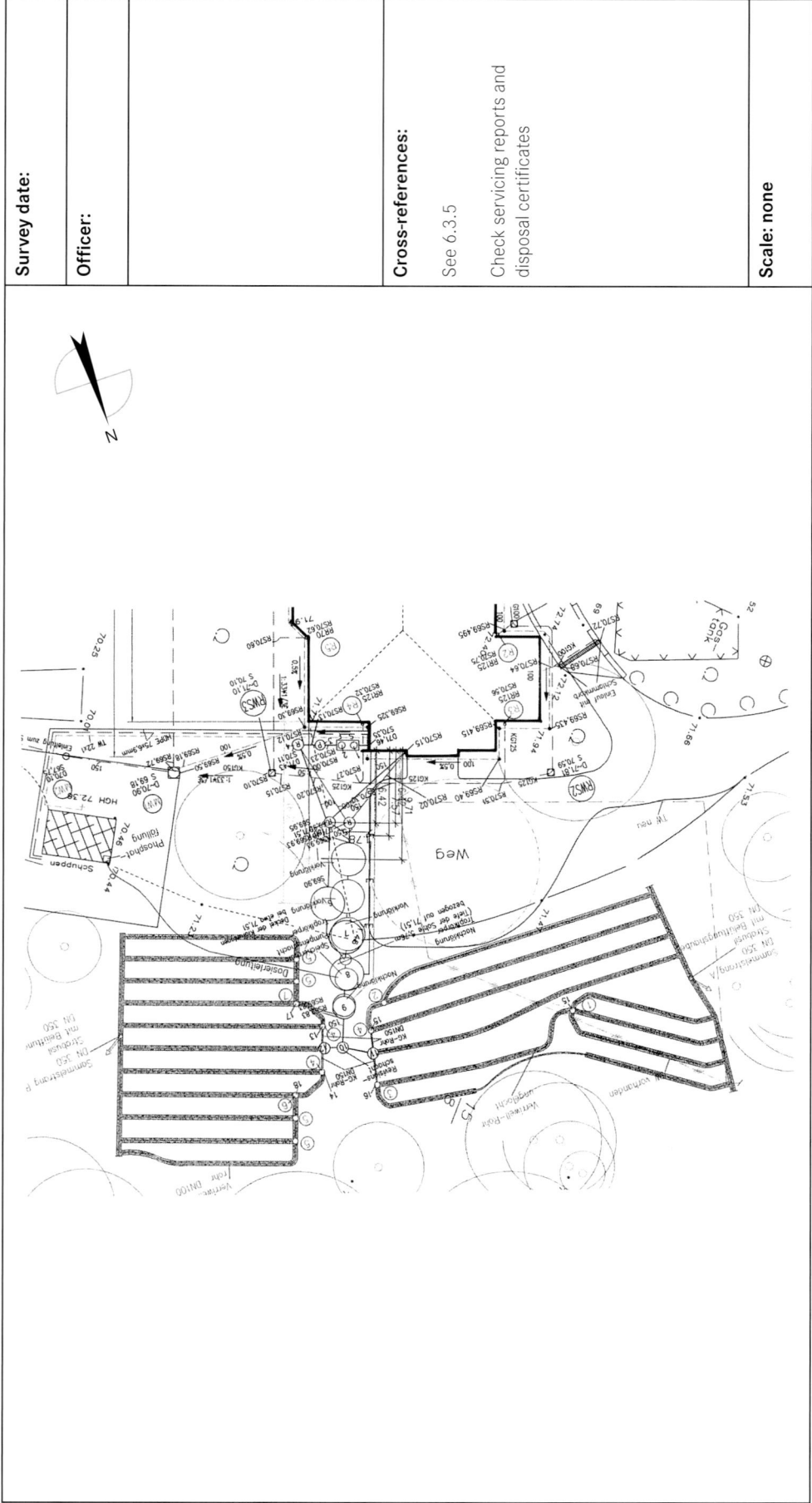

Property: Date: Signature:

Text page / damage record

Element group: Special elements, trickling filter system

Element:			
Findings no.	Description	Necessary action	Initiated date / signature	Completed date / signature

Property:

Date:

Signature:

Condition Survey / building services element group

Code 6.3.8.1

Element: Heating system
Participants:
On-site inspection results regarding issues for clarification, as listed in analysis of preliminary investigations (see Building Maintenance Logbook Code 6.2.2):
Other conspicuous features / defects

Property: Date: Signature:

Building services element group: Heating system
See also maintenance contract

Code 6.3.8.1

Damage type/cause	Findings	n.f.	Damage type/cause	Findings	n.f.	Damage type/cause	Findings	n.f.
Maintenance contract met on schedule			Aesthetic defects					
Maintenance contract met in full			Constructional defects: loose supports					
Comments from servicing report considered			Missing insulation					
System changes documented								
			Other					
System optimization recommended								
Consumption p.a.								
Operational faults/users								
Reliability of maintenance contractor in case of faults/users								

Property: Date: Signature:

Condition Survey / building services element group

Element: Sanitary and drainage system

Participants:

On-site inspection results regarding issues for clarification, as listed in analysis of preliminary investigations (see Building Maintenance Logbook Code 6.2.2):

Other conspicuous features / defects

Property:

Date:

Signature:

Building services element group: Sanitary and drainage system
See also maintenance contract

Damage type / cause	Findings	n.f.	Damage type / cause	Findings	n.f.
Maintenance contract met on schedule			Aesthetic defects		
Maintenance contract met in full			Constructional defects: loose supports		
Comments from servicing report considered			Missing insulation		
System changes documented					
			Other		
System optimization recommended					
Consumption p.a.					
Operational faults / users					
Reliability of maintenance contractor in case of faults / users					
Emptying certificates:					
Grease interceptors					
Septic tank					

Property: Date: Signature:

Condition Survey / building services element group

Element: Ventilation system

Participants:

On-site inspection results regarding issues for clarification, as listed in analysis of preliminary investigations (see Building Maintenance Logbook Code 6.2.2):

Other conspicuous features / defects

Property:

Date:

Signature:

Building services element group: Ventilation system

See also maintenance contract

Damage type / cause	Findings	n.f.	Damage type / cause	Findings	n.f.	Damage type / cause	Findings	n.f.
Maintenance contract met on schedule			Aesthetic defects					
Maintenance contract met in full			Constructional defects: loose supports					
Comments from servicing report considered			Missing insulation					
System changes documented								
			Other					
System optimization recommended								
Operational faults / users								
Reliability of maintenance contractor in case of faults / users								

Property: Date: Signature:

197

Element: Electrical installation

Participants:

On-site inspection results regarding issues for clarification, as listed in analysis of preliminary investigations (see Building Maintenance Logbook Code 6.2.2):

Other conspicuous features/defects

Property:

Date:

Signature:

Building services element group: Electrical installation
See also maintenance contract

Damage type/cause	Findings	n.f.	Damage type/cause	Findings	n.f.	Damage type/cause	Findings	n.f.
Maintenance contract met on schedule			Aesthetic defects					
Maintenance contract met in full			Constructional defects: loose supports					
Comments from servicing report considered			Missing insulation					
System changes documented								
			Other					
System optimization recommended								
Consumption p.a.								
Operational faults/users								
Reliability of maintenance contractor in case of faults/users								

Property: Date: Signature:

Condition Survey/building services element group

Code 6.3.8.5

Element: Lightning protection system

Participants:

On-site inspection results regarding issues for clarification, as listed in analysis of preliminary investigations (see Building Maintenance Logbook Code 6.2.2):

Other conspicuous features/defects

Property:

Date:

Signature:

Building services element group: Lightning protection system
See also maintenance contract

Code 6.3.8.5

Damage type / cause	Findings	n.f.	Damage type / cause	Findings	n.f.	Damage type / cause	Findings	n.f.
Maintenance contract met on schedule			Aesthetic defects					
Maintenance contract met in full			Constructional defects: loose supports					
Comments from servicing report considered			Missing insulation					
System changes documented								
			Other					
System optimization recommended								
Consumption p.a.								
Operational faults / users								
Reliability of maintenance contractor in case of faults / users								

Property: Date: Signature:

Element: Firefighting system / equipment

Participants:

On-site inspection results regarding issues for clarification, as listed in analysis of preliminary investigations (see Building Maintenance Logbook Code 6.2.2):

Other conspicuous features / defects

Property:

Date:

Signature:

Building services element group: Firefighting system / equipment   Code 6.3.8.6

See also maintenance contract

Damage type / cause	Findings	n.f.	Damage type / cause	Findings	Damage type / cause	Findings	n.f.
Maintenance contract met on schedule			Aesthetic defects				
Maintenance contract met in full			Constructional defects: loose supports				
Comments from servicing report considered			Missing insulation				
System changes documented							
			Other				
System optimization recommended							
Consumption p.a.							
Operational faults / users							
Reliability of maintenance contractor in case of faults / users							

Property: Date: Signature:

203

Condition Survey/building services element group

Element: Fire alarm system

Participants:

On-site inspection results regarding issues for clarification, as listed in analysis of preliminary investigations (see Building Maintenance Logbook Code 6.2.2):

Other conspicuous features / defects

Property: Date: Signature:

Building services element group: Fire alarm system   Code 6.3.8.7
See also maintenance contract

Damage type / cause	Findings	n.f.	Damage type / cause	Findings	n.f.	Damage type / cause	Findings	n.f.
Maintenance contract met on schedule			Aesthetic defects					
Maintenance contract met in full			Constructional defects: loose supports					
Comments from servicing report considered			Missing insulation					
System changes documented			Other					
System optimization recommended								
Consumption p.a.								
Operational faults / users								
Reliability of maintenance contractor in case of faults / users								

Property: Date: Signature:

9 Schloss Reichenow

The cultural landscape of the German State of Brandenburg, which surrounds the national capital of Berlin in the east of the country, is renowned not only for its village churches, but also for its 500-plus castles, palaces, manor houses and stately homes. This cultural and artistic heritage lends the region its unmistakable character. However, in the wake of German reunification in 1989, many of these cultural artefacts stood empty or fell into neglect, due in part to the difficult ownership situation, and dilapidation threatened. It is the duty and mission of the Deutsche Stiftung Denkmalschutz (German Foundation for Monument Protection) to preserve this heritage. Consequently, a pilot venture launched in 1992 saw the formation – in Partnership with the Federal State of Brandenburg – of Brandenburgische Schlösser GmbH (BSG), a charitable operating company charged with rescuing selected castles and palaces of special architectural or historic interest.

Schloss Reichenow was among the first heritage buildings to undergo full-scale refurbishment by BSG, prior to reoccupation (Figure 1).

Figure 1:
View of portico
with driveway, 1998

Schloss Reichenow is fairly young as far as stately homes go, built as late as 1897 in the neo-gothic style. The family of the client and owner, Freiherr von Eckardstein, occupied the palace and ran the surrounding estates until the end of the Second World War. After the war, the building first served as accommodation for eastern European refugees and was later a seat of the municipal administration, a residence, a restaurant and a kindergarten. In the late 1980s, it stood largely vacant and remained unattended for a long period. This lack of occupation, compounded by a maintenance backlog built up over many years, gradually led to widespread dilapidation (Figure 2). Schloss Reichenow finally passed into the ownership of BSG in 1994.

Figure 2: View of portico prior to refurbishment, 1995

Figure 3: Portico front after refurbishment, 2010

Figure 4: Dining hall with restored panel parquetry after refurbishment, 1997

Figures 5 and 6: Conference room, 1997

The palace features a contemporary internal circulation system. The straightforward, clearly-structured layout, free from nooks and crannies, reflects the purity of the architectural style and enhances the building's utility. Its suitability as a hotel was apparent even during the early project stages, on account of the manageable size (30 rooms and halls) (Figures 4 and 6), double-loaded central corridor arrangement on each level, and its delightful setting, overlooking a small lake. Yet, as with the other stately homes in the custody of BSG, occupancy was viewed solely as a means of securing long-term preservation of the historic fabric through use. Indeed, the majority of restoration decisions taken by BSG in its various projects have been guided by this principle of maximum conservation.

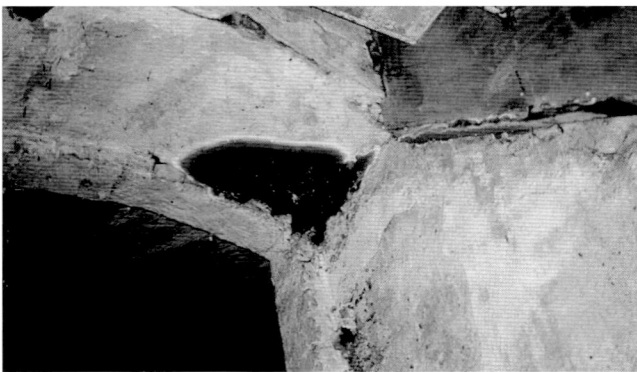

Figure 7: Fruiting bodies of dry rot (boletus destructure)

Figure 8: Spalled rendering, 1995

The first inspections of Schloss Reichenow showed the building to be in a deplorable state. Buckets and bowls were positioned at various points on the attic floor in a makeshift attempt to prevent infiltrating rainwater downpipes, and dry rot (boletus destructor) (Figure 7) fruiting bodies had spread throughout the building, wreaking havoc on roof timbers, wooden beams, window shutters and various other interior fixtures.

Rusting metal sheets, fixed along the lavish crenellations (Figures 9 to 12), testified to a further emergency measure aimed at preventing water ingress, while large patches of render crumbled from the façades below (Figure 8). Here, frost action had also spalled the perimeter neo-gothic pointed-arch frieze to expose the underlying iron supports (Figures 13 to 16). A temporary flat roof (Figure 17) marked the position of the former tower, which had been destroyed during the war. Missing or replaced parapets units, bricked-up window panels (Figures 18 to 20), poorly overpainted fibrous plaster ceilings, and a succession of crudely installed suspended ceilings further disfigured the once majestic palace.

Figure 9: View from roof to eaves, 1995

Figure 10: Park-front balcony over bay window prior to refurbishment, with crenellations and neogothic pointed-arch frieze, 1995

Figure 11: Weathered eaves crenellations and defective rainwater drainage, 1995

Figure 12: Park-front balcony over bay window after refurbishment, 2010

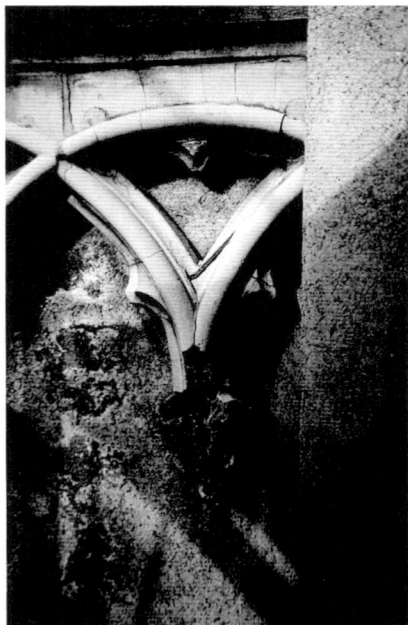

Figure 13: Neogothic pointed-arch frieze prior to refurbishment, 1995

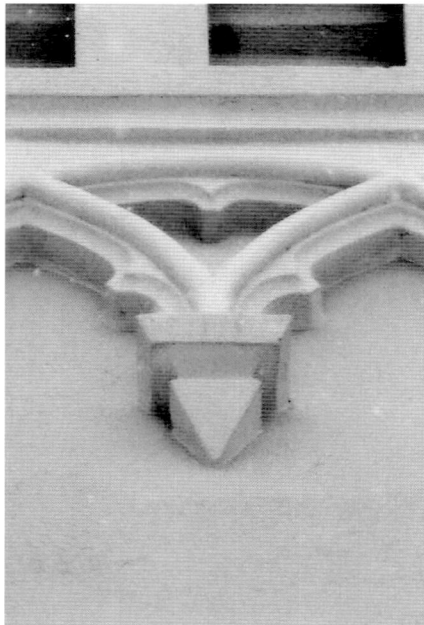

Figure 14: Neogothic pointed-arch frieze after refurbishment, 2010

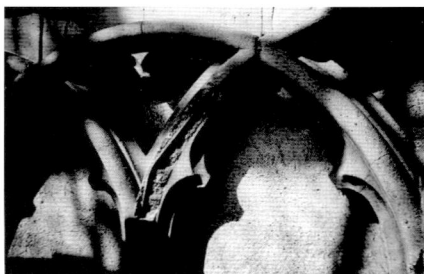

Figures 15 and 16: Details of neogothic pointed-arch frieze prior to refurbishment, 1995

Figure 17: Tower shaft with temporary roofing prior to refurbishment, 1995

Figure 18: Walled-up window opening, 1995

Figures 19 and 20: Window opening after refurbishment, left 1997, above 2010

213

The first step was to prevent any further dilapidation of the building through wholesale refurbishment of the roof and dry rot treatment. Using remains of the historic slate roofing discovered behind ashlaring in the loft, it was possible to track down the source of the large size English slates and procure replacements. The copings, crenellations and pinnacles were repaired stone by stone and weatherproofed using a restrained, though technically reliable, zinc-sheet flashing solution (Figure 21). The eaves drainage installation comprised box gutters, walkable for maintenance, that were carried behind the crenellations. The necessary lightning protection system was discreetly bonded to the metal gutters behind the crenellations so as not to detract from the façade elevation. One of the key restoration issues concerned the treatment of the temporary roofing at the site of the destroyed tower (Figure 17).

Figure 21: Roofscape with English slate after refurbishment, 1997

The former tower had constituted such a powerful articulating feature in the overall composition that the decision was taken, in consultation with the conservation planning authorities, for reconstruction (Figure 22). A painstaking survey of historic images, which served as the basis for the production detailing, paved the way for a largely authentic restoration of the ensemble (Figures 23 and 24). By chance it was learned that shortly before the end of the war, a number of precious decorative sandstone features from the building had been dumped in the nearby lake. Assisted by divers from the German Federal Agency for Technical Relief (Technisches Hilfswerk), a surprising quantity of this architectural ornamentation was salvaged. This was either reinstated at its original location in the building or used as a model for full-scale replicas.

Most of the historic and fibrous plasterwork was preserved, being touched up as needed. Of the historic double windows, made from mahogany-stained pine, only the rot-infected timber members needed replacement. Thanks

Figure 22: Reconstruction of tower, 1997

to elaborate joinery repairs, it was possible to preserve many of the doors, windows and even the wall coverings. The full-scale structural refurbishment of the building interior entailed wide-ranging measures for the temporary protection of the fibrous plasterwork, parquet flooring, and wall and ceiling finishes. A specialist contractor was appointed to rectify the defects in the terrazzo flooring in the entrance hall. Although a number of rot-infected structural members of the main stairway (Figure 25) needed replacement, it was possible to retain most of the well-worn treads and risers as the testament to the building's history (Figures 26 to 28). One intriguing find made during restoration was an 80%-preserved fibrous plaster ceiling beneath a suspended ceiling in the upper hall (Figures 29 to 32).

Removal of a retrofitted drywall partition enclosing the upper stair flight not only uncovered a historic balustrade with neo-gothic decoration (Figure 33 and 34), but allowed reinstatement of the open, ornamental character of the hall (Figure 28). The most far-reaching intervention in the

Figure 23: The variation in height and level of the elements of the Schloss Reichenow - topped by the reconstructed tower - are very much part of the distinctive charm of this refurbished building.

Figure 24: Ornamental friezes and neogothic balustrades feature prominently in the overall composition, 1997

Figure 25: Stair destroyed by dry rot, though with decorative neogothic balustrade still mounted on stair string

Figure 26-28: Stairway (first floor) and hall revelling in their full historic splendour after refurbishment, 1997

Figure 29: Neogothic balustrade,
hidden for years behind drywall partition

Figure 30 and 31: Rediscovered historic fibrous
plasterwork in first-floor hall, hidden behind
suspended ceiling, 1996

Figure 32: Ceiling in first-floor hall
and neogothic balustrade after refurbishment, 2010

Figure 33: Neogothic balustrade
after being uncovered, 1996

Figure 34: Neogothic balustrade
after refurbishment, 2010

building fabric was needed for the bathrooms in the hotel rooms. Fortunately, it was possible to house all the necessary supply and extract ventilation installations in the basement, with the supply air openings concealed in the basement windows behind the ramped driveway. Extracted air is discreetly discharged above roof level via the existing chimneys. The historic panel parquetry in the garden hall (Figure 4) was newly laid while, elsewhere, the historic wood-board and parquet flooring was largely conserved. The gardens, of approximately five hectares, were restored to their original form. The palace parkland and nearby lake once again blended together in a natural harmony amid the captivating landscape (Figure 35).

Figure 35: Veranda overlooking park with visual links to nearby lake, 1998

Following a three-year construction period at a total cost of around € 5.7 million, Schloss Reichenow was ceremoniously inaugurated as a hotel in 1997. Today, the palace and hotel boast a reputation extending far beyond the Berlin and Brandenburg region. The building even occasionally hosts several wedding celebrations on a single day.

Our avowed aim, upon successful completion of the restoration works, was to ensure the long-term maintenance of the heritage building and its fabric in a high-quality condition. The Building Maintenance Logbook is the key tool for this purpose.

Schloss Reichenow served as the basis for the specimen User Maintenance Instructions (Section 7) and Condition Survey spreadsheets (Section 8) presented in this book.